Comparative Approaches to Using R and Python for Statistical Data Analysis

Rui Sarmento
University of Porto, Portugal

Vera Costa
University of Porto, Portugal

A volume in the Advances in Systems Analysis, Software Engineering, and High Performance Computing (ASASEHPC) Book Series

IGI GLOBAL
DISSEMINATOR OF KNOWLEDGE

www.igi-global.com

Published in the United States of America by
IGI Global
Information Science Reference (an imprint of IGI Global)
701 E. Chocolate Avenue
Hershey PA 17033
Tel: 717-533-8845
Fax: 717-533-8661
E-mail: cust@igi-global.com
Web site: http://www.igi-global.com

Library of Congress Cataloging-in-Publication Data

Names: Sarmento, Rui, 1979- | Costa, Vera, 1983-
Title: Comparative approaches to using R and Python for statistical data analysis / by Rui Sarmento and Vera Costa.
Description: Hershey PA : Information Science Reference, [2017] | Includes bibliographical references and index.
Identifiers: LCCN 2016050989| ISBN 9781683180166 (hardcover) | ISBN 9781522519898 (ebook)
Subjects: LCSH: Mathematical statistics--Data processing. | R (Computer program language) | Python (Computer program language)
Classification: LCC QA276.45.R3 S27 2017 | DDC 519.50285/5133--dc23 LC record available at https://lccn.loc.gov/2016050989

This book is published in the IGI Global book series Advances in Systems Analysis, Software Engineering, and High Performance Computing (ASASEHPC) (ISSN: 2327-3453; eISSN: 2327-3461)

British Cataloguing in Publication Data
A Cataloguing in Publication record for this book is available from the British Library.

Advances in Systems Analysis, Software Engineering, and High Performance Computing (ASASEHPC) Book Series

ISSN:2327-3453
EISSN:2327-3461

Editor-in-Chief: Vijayan Sugumaran, Oakland University, USA

Mission

The theory and practice of computing applications and distributed systems has emerged as one of the key areas of research driving innovations in business, engineering, and science. The fields of software engineering, systems analysis, and high performance computing offer a wide range of applications and solutions in solving computational problems for any modern organization.

The **Advances in Systems Analysis, Software Engineering, and High Performance Computing (ASASEHPC) Book Series** brings together research in the areas of distributed computing, systems and software engineering, high performance computing, and service science. This collection of publications is useful for academics, researchers, and practitioners seeking the latest practices and knowledge in this field.

Coverage

- Performance Modelling
- Computer System Analysis
- Computer Networking
- Engineering Environments
- Human-Computer Interaction
- Metadata and Semantic Web
- Software Engineering
- Distributed Cloud Computing
- Enterprise Information Systems
- Virtual Data Systems

IGI Global is currently accepting manuscripts for publication within this series. To submit a proposal for a volume in this series, please contact our Acquisition Editors at Acquisitions@igi-global.com or visit: http://www.igi-global.com/publish/.

The Advances in Systems Analysis, Software Engineering, and High Performance Computing (ASASEHPC) Book Series (ISSN 2327-3453) is published by IGI Global, 701 E. Chocolate Avenue, Hershey, PA 17033-1240, USA, www.igi-global.com. This series is composed of titles available for purchase individually; each title is edited to be contextually exclusive from any other title within the series. For pricing and ordering information please visit http://www.igi-global.com/book-series/advances-systems-analysis-software-engineering/73689. Postmaster: Send all address changes to above address.

Titles in this Series

For a list of additional titles in this series, please visit:
http://www.igi-global.com/book-series/advances-systems-analysis-software-engineering/73689

Resource Management and Efficiency in Cloud Computing Environments
Ashok Kumar Turuk (National Institute of Technology Rourkela, India) Bibhudatta Sahoo (National Institute of Technology Rourkela, India) and Sourav Kanti Addya (National Institute of Technology Rourkela, India)
Information Science Reference • ©2017 • 352pp • H/C (ISBN: 9781522517214) • US $205.00

Handbook of Research on End-to-End Cloud Computing Architecture Design
Jianwen "Wendy" Chen (IBM, Australia) Yan Zhang (Western Sydney University, Australia) and Ron Gottschalk (IBM, Australia)
Information Science Reference • ©2017 • 507pp • H/C (ISBN: 9781522507598) • US $325.00

Innovative Research and Applications in Next-Generation High Performance Computing
Qusay F. Hassan (Mansoura University, Egypt)
Information Science Reference • ©2016 • 488pp • H/C (ISBN: 9781522502876) • US $205.00

Developing Interoperable and Federated Cloud Architecture
Gabor Kecskemeti (University of Miskolc, Hungary) Attila Kertesz (University of Szeged, Hungary) and Zsolt Nemeth (MTA SZTAKI, Hungary)
Information Science Reference • ©2016 • 398pp • H/C (ISBN: 9781522501534) • US $210.00

Managing Big Data in Cloud Computing Environments
Zongmin Ma (Nanjing University of Aeronautics and Astronautics, China)
Information Science Reference • ©2016 • 314pp • H/C (ISBN: 9781466698345) • US $195.00

Emerging Innovations in Agile Software Development
Imran Ghani (Universiti Teknologi Malaysia, Malaysia) Dayang Norhayati Abang Jawawi (Universiti Teknologi Malaysia, Malaysia) Siva Dorairaj (Software Education, New Zealand) and Ahmed Sidky (ICAgile, USA)
Information Science Reference • ©2016 • 323pp • H/C (ISBN: 9781466698581) • US $205.00

For an enitre list of titles in this series, please visit:
http://www.igi-global.com/book-series/advances-systems-analysis-software-engineering/73689

IGI GLOBAL
DISSEMINATOR OF KNOWLEDGE

www.igi-global.com

701 East Chocolate Avenue, Hershey, PA 17033, USA
Tel: 717-533-8845 x100 • Fax: 717-533-8661
E-Mail: cust@igi-global.com • www.igi-global.com

To our parents and family…

Table of Contents

Preface

We may at once admit that any inference from the particular to the general must be attended with some degree of uncertainty, but this is not the same as to admit that such inference cannot be absolutely rigorous, for the nature and degree of the uncertainty may itself be capable of rigorous expression. – Sir Ronald Fisher

The importance of Statistics in our world is increasing greatly in recent decades. Due do the need to provide inference from data samples; statistics is one of the greatest achievements of humanity. Its use has spread to a large range of research areas, not only limited to research done by mathematicians or pure statistics professionals. Nowadays, it is standard procedure to include some statistical analysis when the scientific study involves data. There is a high influence and demand for statistical analysis in today's Medicine, Biology, Psychology, Physics and many other areas.

The demand for statistical analysis of data has proliferated so much; it has survived inclusively to attacks from the mathematical challenged.

If the statistics are boring, then you've got the wrong numbers. – Edward R. Tufte

Thus, with the advent of computers and advanced computer software, the intuitiveness of analysis software has evolved greatly in recent years and they have opened to a wider audience of users. It is common to see another kind of statistical researchers in modern academies. Those with no advanced studies in the mathematical areas are the new statisticians and use and produce statistical studies with scarce or no help from others.

Above all else show the data. – Edward R. Tufte

The need to expose the studies in a clear fashion for a non-specialized audience has brought the development of, not only intuitive software but software directed to the visualization of data and data analysis. For example, the psychologist with no mathematical foundations can now choose from several languages and software to add value to their studies by performing throughout analysis of their data and present it in an understandable fashion.

This book presents a comparison of two of the available languages to execute data analysis and statistical analysis, R language and also the Python language. It is directed to anyone, experienced or not, that might need to analyze his/her data in an understandable way. For those more experienced, the authors of this book approach the theoretical fundamentals of statistics, and for a larger range of audience, explain the programming fundamentals, both with R and Python languages.

The statistical tasks range from Descriptive Analytics. The authors describe the need for basic statistical metrics and present the main procedures with both languages. Then, Inferential Statistics are presented in this book. High importance is given to the most needed statistical tests to perform a coherent data analysis. Following Inferential Statistics, the authors also provide examples, with both languages, in a throughout explanation of Factor Analysis. The authors emphasize the importance of variable study and not only the objects study. Nonetheless, the authors present a chapter also dedicated to the clustering analysis of studied objects. Finally, an introductory study of regression models and linear regression is also tabled in this book.

The authors do not deny that the structure of the book might pose some comparison questions since the book deals with two different programming languages. The authors end the book with a discussion that provides some clarification on this subject but, above all, also provides some insights for further consideration.

Finally, the authors would like to thank all the colleagues that provided suggestions and reviewed the manuscript in all its development phases, and all the friends and family members for their support.

Introduction

TECHNOLOGY AND CONTEXT INTEGRATION

This book enables the understanding of procedures to execute data analysis with the Python and R languages. It includes several reference practical exercises with sample data. These examples are distributed in several statistical topics of research, ranging from easy to advanced. The procedures are throughout explained and are comprehensible enough to be used by non-statisticians or data analysts. By providing the solved tests with R and Python, the proceedings are also directed to programmers and advanced users. Thus, the audience is quite vast, and the book will fulfill either the curious analyst or the expert.

At the beginning, we explain who is this book for and what the audience gains by exploring this book. Then, we proceed and explain the technology context by introducing the tools we use in this book. Additionally, we present a summarizing diagram with a workflow appropriated for any statistical data analysis. At the end, the reader will have some knowledge of the origins and features of the tools/languages and will be prepared for further reading of the subsequent chapters.

WHO IS THIS BOOK FOR?

This book mainly solves the problem of a broad audience not oriented to mathematics or statistics. Nowadays, many human sciences researchers need to do the analysis of their data with few or no knowledge about statistics. Additionally, they have even less knowledge of how to use necessary tools for the task. Tools like Python and R, for example. The uniqueness of this book is that it includes procedures for data analysis from pre-processing to final results, for both Python and R languages. Thus, depending on the knowledge level or the needs of the reader it might be very compelling to choose one or another tool to solve the problem. The authors believe both tools have their advantages and disadvantages when compared to each other, and those are outlined in this book. Succinctly, this book is appropriated for:

- End users of applications and both languages,
- Undergraduate/Graduate Students,
- Human Sciences Professionals,
- Marketing Specialists,
- Data Analysts,
- Statisticians.

This broad audience will benefit from reading this book and will better use the tools. They will be able to approach their data analysis problems with better understanding of data analysis and the recommended tools to execute these tasks.

TECHNOLOGY CONTEXT

This book provides a very detailed approach to statistical areas. First, we introduce Python and R to the reader. The uniqueness of this book is that it provides a way the reader can feel motivated to experiment with one of the languages or even both. As a bonus, both languages have an inherent flexibility as programming languages. This is an advantage when compared to "what-you-see-is-what-you-get" solutions as SPSS or others.

Tools

There are many information sources about these two languages. We will state a brief summary about both languages origin. This information source ranges from the language authors themselves to several blogs available on the World Wide Web.

R

Ross Ihaka and Robert Gentleman conceived R Language with most of its influences from the S language conceived by Rick Becker and John Chambers. There were several features R author's thought could be added to S (Ihaka & Gentleman 1996).

The R language authors worked at the University of Auckland and had an interest in statistical computing but felt there were limitations in the offering of these types of solutions in their Macintosh laboratory. The authors felt a suitable commercial environment didn't yet exist and they began to experiment and to develop one.

Despite the similarity between R and S, some fundamental differences remain, according to the language authors (Ihaka & Gentleman 1998):

Memory Management: In R, we allocate a fixed amount of memory at startup and manage it with an on-the-fly garbage collector. This means that there is tiny heap growth and as a result there are fewer paging problems than are seen in S.
Scoping: In S, variables in functions are either local or global. In R, we allow functions to access to the variables which were in effect when the function was defined; an idea which dates back to Algol 60 and found in Scheme and other lexically scoped languages. In S, the variable being manipulated is global. In R, it is the one which is in effect when the function is defined; i.e. it is the argument to the function itself. The effect is to create a variable which only the inner function can see and manipulate.
The scoping rules used in R have met with approval because they promote a very clean programming style. We have retained them despite the fact that they complicate the implementation of the interpreter.

As the authors emphasize, scoping in R provides a cleaner way to program despite the fact that it complicates the needed code interpretation. As we will see throughout the book, this R feature makes way for R being a very clean and intuitive language, which facilitates coding even without previous programming experience.

The authors continue and explain other differences to previous attempts to build a statistical programming language:
The two differences noted above are of a very basic nature. Also, we have experimented with some other features in R. A good deal of the experimentation has been with the graphics system (which is quite similar to that of S). Here is a brief summary of some of these experiments.
Colour Model: R uses a device independent 24-bit model for color graphics. Colors can be specified in some ways.

1. *By defining the levels of red, green and blue primaries, which make up the Colour. For example, the string "#FFFF00" indicates full intensity for red and green with no blue; producing yellow.*
2. *By giving a color name. R uses the color naming system of the X Window System to provide about 650 standard color names, ranging from the plain "red", "green" and "blue" to the more exotic "light goldenrod", and "medium orchid 4".*

3. *As an index into a user settable color table. This provides compatibility with the S graphics system.*

Line Texture Description: Line textures can also be specified in a flexible fashion. The specification can be:

1. *A texture name (e.g. "dotted").*
2. *A string containing the lengths for the pen up/down segments which compose a line. For example, the specification "52" indicates 5 points (or pixels) with "pen down" followed by 2 with "pen up", with the pattern replicated for the length of the line.*
3. *An index into a fixed set of line types, again providing compatibility with S.*

From the previous statements, the reader should already notice that there is much importance given by the authors to the need to customize the optical output of the statistical data analysis. This feature is also an important R language characteristic and helps the user to achieve good visual outputs.

Regarding mathematical features, the authors continue and describe some more features yet:

Mathematical Annotation: Paul Murrell and I have been working on a simple way of producing mathematical annotation in plots. Mathematical annotation is provided by specifying an unevaluated R expression instead of a character string. For example, expression (x^2+1) can be used to produce the mathematical expression x^2+1 as annotation in a plot.
The annotation system is fairly straightforward, and not designed to have the full capabilities of a system such as TeX. Even so, it can produce quite nice results.

From the previous authors' statements, high versatility in the mathematical annotation of graphs, plots, and charts is expected. The authors compare this lower complexity to another language, TeX language which is frequently used by the researchers when they need to produce scientific literature. This way, the authors expect the user to, for example, create a plot with a single R command which itself uses an expression to describe the labels with mathematical notation.

The authors then again continue with the explanation about R, and more specifically, about plots:

Flexible Plot Layouts: A part of his Ph.D. research, Paul Murrell has been looking at a scheme for specifying plot layouts. The scheme provides a simple way of determining how the surface of the graphs device should be divided up into some rectangular plotting regions. The regions can be constrained in a variety of ways. Paul's original work was in Lisp, but he has implemented a useful subset to R.
These graphical experiments were carried out at Auckland, but others have also bound R to be an environment which can be used as a base for experimentation.

Thus, R language, as introduced here by the authors themselves, provides a very strong bond with the user by being masterfully customizable and focused on the excellence of the visual output.

Python

Regarding Python, its history started back in the 20th century. The following summary about Python is available from Wikipedia and several web pages where some significant milestones in the development of the language have been written.

Guido van Rossum at CWI in the Netherlands first idealized the Python programming language in the late 1980s.

Python was conceived at the end of the1980s (Venners 2003), and its implementation was started in December 1989 (van Rossum 2009) as a successor to the ABC programming language, capable of exception handling and interfacing with the Amoeba operating system (van Rossum 2007).

Python is said to have several influences from other programming languages too. Python's core syntax and some aspects of its construction are indeed very similar to ABC. Other languages also provided some of Python's syntax like, for example, C. Regarding the followed model for the interpreter, which becomes interactive when running without arguments, the authors borrowed from the Bourne shell case study. Python regular expressions, for example, used for string manipulation, where derived from Perl language (Foundation 2007b).

Python Version 2.0 was released on October 16, 2000, with many major new features including better memory management. However, the most remarkable change was the development process itself, an agiler and depending on a community of developers, enabling a process depending on network efforts (Kuchling & Zadka 2009).

Python's standard library additions and syntactical choices were also strongly influenced by Java in some cases. Examples of such additions to the library were, for instance:

- The logging package introduced in version 2.3 (Kuchling 2009, Sajip & Mick 2002).
- The threading package for multithreaded applications.
- The SAX parser, introduced in 2.0, and the decorator syntax that uses @, was made available from version 2.4 (Foundation 2007c, Smith, Jewett, Montanaro & Baxter 2003).
- Another example of these java influenced libraries, Python's method resolution order was changed in Python 2.3 to use the C3 linearization algorithm as employed in Dylan programming language (Foundation 2007a).

Python is currently in version 3.x and the main characteristics of this release are:

- Python 3.0, a major, backwards-incompatible release, was published on December 3, 2008(Foundation 2008) after an extended period of testing. Many of its major features have also been backported to the backwards-compatible Python 2.6 and 2.7 (van Rossum 2006).
- Python 3.0 was developed with the same philosophy as in prior versions. However, as Python had accumulated new and redundant ways to program the same task, Python 3.0 had an emphasis on removing duplicative constructs and modules. Nonetheless, Python 3.0 remained a multi-paradigm language. Coders still had options among object-orientation, structured programming, and functional programming. However, as it is inherently a multi-paradigm language, Python 3.0 details were more prominent than they were in Python 2.x.

Resuming, Python is a versatile language, depending not on a team of developers but on a community, which, as we will see later in this book, provides several packages that are directed to specific goals. Regarding mathematical and statistics tasks, there are several packages already proposed by the developers' community.

Figure 1. Book map

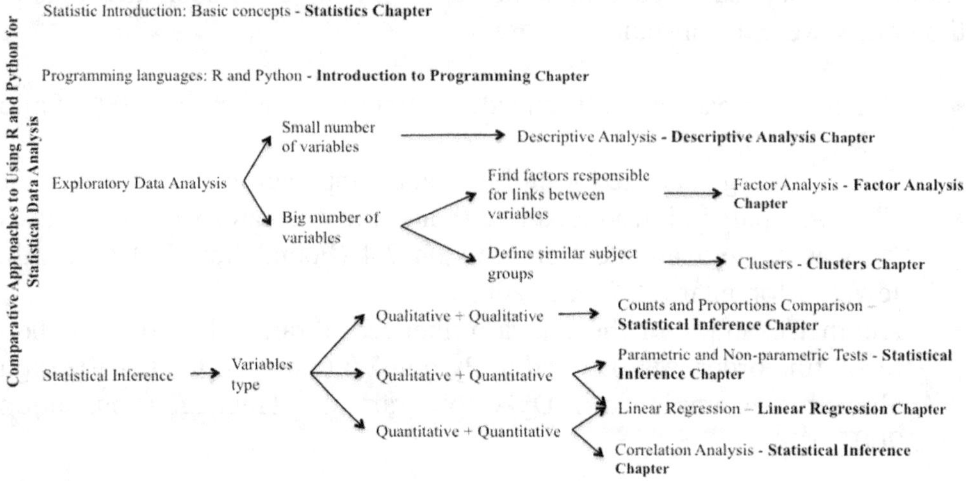

BOOK MAP

The statistical data analysis tasks presented in this book are spread within several chapters. To do a complete data analysis of the data, the reader might have to explore several or all chapters. Nonetheless, if some particular task is needed, the reader might find the workflow diagram in Figure 1 useful. Thus, a decision of which method to use is simplified to the reader, taking account the goal of his/her analysis.

CONCLUSION

This preface presents an introduction and contextualization to the reader of this book. Moreover, a technology context is provided regarding the tools available for the reader to reach his analysis goals. Although the book is organized with a crescent complexity of materials, the reader will encounter an imminently practical book with examples from beginning to end. Nevertheless, the authors of this book will not forget a theoretical introduction to statistics.

Additionally, in this preface, we provided a summary of the birth of the languages we are focusing on this book. We introduced the reader to their creators, and we provide additional literature for the curious readers to explore. Both languages have a community of developers, which provides great speed in the improvement of the languages and the appeasement of new packages and libraries.

Interestingly, although R seems at this point directed to a specific statistics area, it is sufficiently generic and versatile to be considered a language where you can program anything in any possible area. On the other side, we have Python which is apparently a generic programming language, not specifically directed to statistics but which depends on a community of aficionados that produce specific packages directed to a variety of areas including statistics.

REFERENCES

Foundation, P. S. (2007a). *5 pep 318: Decorators for functions and methods*. Retrieved from https://docs.python.org/release/2.4/whatsnew/node6.html

Foundation, P. S. (2007b). *Regular expression operations*. Retrieved from https://docs.python.org/2/library/re.html

Foundation, P. S. (2007c). *Threading — Higher-level threading interface*. Retrieved from https://docs.python.org/2/library/threading.html

Foundation, P. S. (2008). *Python 3.0 release*. Retrieved from https://www.python.org/download/releases/3.0/

Foundation, P. S. (n.d.). *8 pep 282: The logging package*. Retrieved from https://docs.python.org/release/2.3/whatsnew/node9.html

Ihaka, R., & Gentleman, R. (1996). R: A language for data analysis and graphics. *Journal of Computational and Graphical Statistics*, *5*, 299–314.

Ihaka, R., & Gentleman, R. (1998). *Genesis*. Retrieved from https://cran.r-project.org/doc/html/interface98-paper/paper_1.html

Kuchling, A. (2009). *Regular expression howto*. Retrieved from https://docs.python.org/2/howto/regex.html

Kuchling, A., & Zadka, M. (2009). *What's new in python 2.0*. Retrieved from http://web.archive.org/web/20091214142515/http://www.amk.ca/python/2.0

Sajip, V., & Mick, T. (2002). *Pep 282 – A logging system*. Retrieved from https://www.python.org/dev/peps/pep-0282/

Smith, K. D., Jewett, J. J., Montanaro, S., & Baxter, A. (2003). *Pep 318 – Decorators for functions and methods*. Retrieved from https://www.python.org/dev/peps/pep-0318/

van Rossum, G. (2006). *Pep 3000 – Python 3000.* Retrieved from https://www.python.org/dev/peps/pep- 3000/

van Rossum, G. (2007). *Why was python created in the first place?.* Retrieved from https://docs.python.org/2/faq/general.htmlwhy-was-python-created-in-the-first-place

van Rossum, G. (2009). *The history of python - A brief timeline of python.* Retrieved from http://python-history.blogspot.pt/2009/01/brief-timeline-of-python.html

Venners, B. (2003). *The making of python - A conversation with Guido van Rossum, part I.* Retrieved from http://www.artima.com/intv/pythonP.html

Chapter 1
Statistics

INTRODUCTION

Statistics is a set of methods used to analyze data. The statistic is present in all areas of science involving the collection, handling and sorting of data, given the insight of a particular phenomenon and the possibility that, from that knowledge, inferring possible new results. One of the goals with statistics is to extract information from data to get a better understanding of the situations they represent. Thus, the statistics can be thought of as the science of learning from data.

Currently, the high competitiveness in search technologies and markets has caused a constant race for the information. This is a growing and irreversible trend. Learning from data is one of the most critical challenges of the information age in which we live. In general, we can say that statistic based on the theory of probability, provides techniques and methods for data analysis, which help the decision-making process in various problems where there is uncertainty.

This chapter presents the main concepts used in statistics, and that will contribute to understanding the analysis presented throughout this book.

VARIABLES, POPULATION, AND SAMPLES

In statistical analysis, "variable" is the common characteristic of all elements of the sample or population to which is possible to attribute a number or category. The values of the variables vary from element to element.

DOI: 10.4018/978-1-68318-016-6.ch001

Types of Variables

Statistical variables can be classified as categorical variables or numerical variables.

Categorical variables have values that describe a "quality" or "characteristic" of a data unit, like "which type" or "which category". Categorical variables fall into mutually exclusive (in one category or another) and exhaustive (include several possible options) categories. Therefore, categorical variables are qualitative variables and tend to be represented by a non-numeric value. Categorical variables may be further described as (Marôco, 2011):

- **Nominal:** The data consist of categories only. The variables are measured in discrete classes, and it is not possible to establish any qualification or ordering. Standard mathematical operations (addition, subtraction, multiplication, and division) are not defined when applied to this type of variable. Gender (male or female) and colors (blue, red or green) are two examples of nominal variables.
- **Ordinal:** The data consist of categories that can be arranged in some exact order according to their relative size or quality, but cannot be quantified. Standard mathematical operations (addition, subtraction, multiplication, and division) are not defined when applied to this type of variable. For example, social class (upper, middle and lower) and education (elementary, medium and high) are two examples of ordinal variables. Likert scales (1-"Strongly Disagree", 2-"Disagree", 3-"Undecided", 4-"Agree", 5-"Strongly Agree") are ordinal scales commonly used in social sciences.

Numerical variables have values that describe a measurable quantity as a number, like "how many" or "how much". Therefore, numeric variables are quantitative variables. Numeric variables may be further described as:

- **Discrete:** The data is numerical. Observations can take a value based on a count of a set of distinct integer values. A discrete variable cannot take the value of a fraction of one value and the next closest value. The number of registered cars, the number of business locations, and the number of children in a family, all of which measured as whole units (i.e. 1, 2, or 3 cars) are some examples of discrete variables.
- **Continuous:** The data is numerical. Observations can take any value between a particular set of real numbers. The value given to one observation for a continuous variable can include values as precise as

possible with the instrument of measurement. Height and time are two examples of continuous variables.

Population and Samples

Population

The population is the total of all the individuals who have certain characteristics and are of interest to a researcher. Community college students, racecar drivers, teachers, and college-level athletes can all be considered populations.

It is not always convenient or possible to examine every member of an entire population. For example, it is not practical to ask all students which color they like. However, it is possible, to ask the students of three schools the preferred color. This subset of the population is called a sample.

Samples

A sample is a subset of the population. The reason for the sample's importance is because in many models of scientific research, it is impossible (from both a strategic and a resource perspective) the study of all members of a population for a research project. It just costs too much and takes too much time. Instead, a selected few participants (who make up the sample) are chosen to ensure the sample is representative of the population. And, if this happens, the results from the sample could be inferred to the population, which is precisely the purpose of inferential statistics; using information on a smaller group of participants makes it possible to understand to all population.

There are many types of samples, including:

- A random sample,
- A stratified sample,
- A convenience sample.

They all have the goal to accurately obtain a smaller subset of the larger set of total participants, such that the smaller subset is representative of the larger set.

Independent and Paired Samples

The relationship or absence of the relationship between the elements of one or more samples defines another factor of classification of the sample, particularly important in statistical inference. If there is no type of relationship between the elements of the samples, it is called independent samples. Thus, the theoretical probability of a given subject belonging to more than one sample is null.

On the opposite, if the same subject composes the samples based on some unifying criteria (for example, samples in which the same variable are measured before and after specific treatment on the same subject), it is called paired samples. In such samples, the subjects who are purposely tested are related. It can even be the same subject (e.g., repeated measurements) or subject with paired characteristics (in statistical blocks studies).

DESCRIPTIVE STATISTICS

Descriptive statistics are used to describe the essential features of the data in a study. It provides simple summaries about the sample and the measures. Together with simple graphics analysis, it forms the basis of virtually every quantitative analysis of data. Descriptive statistics allows presenting quantitative descriptions in a convenient way. In a research study, it may have lots of measures. Or it may measure a significant number of people on any measure. Descriptive statistics helps to simplify large amounts of data in a sensible way. Each descriptive statistic reduces lots of data into a simpler summary.

Frequency Distributions

Frequency distributions are visual displays that organize and present frequency counts (n) so that the information can be interpreted more easily. Along with the frequency counts, it may include relative frequency, cumulative frequency, and cumulative relative frequencies.

- The frequency (n) is the number of times a particular variable assumes that value.
- The cumulative frequency (N) is the number of times a variable takes on a value less than or equal to this value.
- The relative frequency (f) is the percentage of the frequency.
- The cumulative relative frequency (F) is the percentage of the cumulative frequency.

Depending on the variable (categorical, discrete or continuous), various frequency tables can be created. See Tables 1 through 6.

Statistics

Table 1. Example 1: favorite color of 10 individuals - categorical variable: list of responses

Blue	Red	Blue	White	Green
White	Blue	Red	Blue	Black

Table 2. Example 1: favorite color of 10 individuals - categorical variable: frequency distribution

Color	n	N	f	F
Blue	4	4	0.4	0.4
Red	2	6	0.2	0.6
White	2	8	0.2	0.8
Green	1	9	0.1	0.9
Black	1	10	0.1	1.0
Total	10		1	

Table 3. Example 2: age of 20 individuals - discrete numerical variable: list of responses

20	22	21	24	21	20	20	24	22	20
22	24	21	25	20	23	22	23	21	20

Table 4. Example 2: age of 20 individuals - discrete numerical variable: frequency distribution

Age	n	N	f	F
20	6	6	0.3	0.3
21	4	10	0.2	0.5
22	4	14	0.2	0.7
23	2	16	0.1	0.8
24	3	19	0.15	0.95
25	1	20	0.05	1
Total	20		1	

Table 5. Example 3: height of 20 individuals - continuous numerical variable: list of responses

1.58	1.56	1.77	1.59	1.63	1.58	1.82	1.69	1.76	1.60
1.73	1.51	1.54	1.61	1.67	1.72	1.75	1.55	1.68	1.65

Table 6. Example 3: height of 20 individuals - continuous numerical variable: frequency distribution

Interval	n	N	f	F
]1.50, 1.55]	3	3	0.15	0.15
]1.55, 1.60]	5	8	0.25	0.4
]1.60, 1.65]	3	11	0.15	0.55
]1.65, 1.70]	3	14	0.15	0.7
]1.70, 1.75]	3	17	0.15	0.85
]1.75, 1.80]	2	19	0.1	0.95
]1.80, 1.85]	1	20	0.05	1
Total	20		1	

Measures of Central Tendency and Measures of Variability

A measure of central tendency is a numerical value that describes a data set, by attempting to provide a "central" or "typical" value of the data (McCune, 2010). As such, measures of central tendency are sometimes called measures of central location. They are also classed as summary statistics.

Measures of central tendency should have the same units as those of the data values from which they are determined. If no units are specified for the data values, no units are specified for the measures of central tendency.

The mean (often called the average) is most likely the measure of central tendency that the reader is most familiar with, but there are others, such as the median, the mode, percentiles, and quartiles.

The mean, median and mode are all valid measures of central tendency, but under different conditions, some measures of central tendency become more appropriate to use than others.

A measure of variability is a value that describes the spread or dispersion of a data set to its central value (McCune, 2010). If the values of measures of variability are high, it signifies that scores or values in the data set are widely spread out and not tightly centered on the mean. There are three common measures of variability: the range, standard deviation, and variance.

Mean

The mean (or average) is the most popular and well-known measure of central tendency. It can be used with both discrete and continuous data. An important property of the mean is that it includes every value in the data set as part of the calculation. The mean is equal to the sum of all the values of the variable divided by the number of values in the data set. So, if we have *n* values in a data set and $(x_1, x_2, \ldots, x_n)$ are values of the variable, the sample mean, usually denoted by $\overline{x}$ (denoted by μ, for population mean), is:

$$\overline{x} = \frac{x_1 + x_2 + \ldots + x_n}{n} = \frac{\sum_{i=1}^{n} x_i}{n}$$

Applying this formula to example 2 above, the mean is given by:

$$\overline{x} = \frac{20 * 6 + 21 * 4 + 22 * 4 + 23 * 2 + 24 * 3 + 25 * 1}{20} = \frac{435}{20} = 21.75$$

So, the age mean for the 20 individuals is around 22 years (approximately).

Median

The median is the middle value or the arithmetic average of the two middle values of the variable that has been arranged in order of magnitude. So, 50% of the observations are greater or equal to the median, and 50% are less or equal to the median. It should be used with ordinal data. The median (after ordering all values) is as follows:

$$\tilde{x} = \begin{cases} \dfrac{x_{\frac{n}{2}} + x_{\frac{n+1}{2}}}{2}, & if\ n\ is\ even \\ x_{\frac{n+1}{2}}, & if\ n\ is\ odd \end{cases}$$

In example 2 above, by ordering the age variable values, we have:

20, 20, 20, 20, 20, 20, 21, 21, 21, **21, 22**, 22, 22, 22, 23, 23, 24, 24, 24, 25

As n is even, the median is the average of the middle values. So $\tilde{x} = \dfrac{21+22}{2} = 21.5$ is the age median for the sample with 20 individuals.

Mode

The mode is the most common value (or values) of the variable. A variable in which each data value occurs the same number of times has no mode. If only one value occurs with the greatest frequency, the variable is unimodal; that is, it has one mode. If exactly two values occur with the same frequency, and that is higher than the others, the variable is bimodal; that is, it has two modes. If more than two data values occur with the same frequency, and that is greater than the others, the variable is multimodal; that is, it has more than two modes (McCune, 2010). The mode should be used only with discrete variables.

In example 2 above, the most frequent value of age variable is "20". It occurs six times. So, "20" is the mode of the age variable.

Percentiles and Quartiles

The most common way to report relative standing of a number within a data set is by using percentiles (Rumsey, 2010). The P_{th} percentile cuts the data set in two so that approximately P% of the data is below it and (100−P)% of the data is above it. So, the percentile of order p is calculated by (Marôco, 2011):

$$P_p = \begin{cases} X_{int(i+1)} & if\ i = \dfrac{np}{100}\ is\ not\ integer \\ \dfrac{X_i + X_{i+1}}{2} & if\ i = \dfrac{np}{100}\ is\ integer \end{cases}$$

where n is the sample size and $int(i+1)$ is the integer part of $i+1$.

It is usual to calculate the P_{25} also called first quartile (Q_1), P_{50} as second quartile (Q_2) or median and P_{75} as the third quartile (Q_3).

In example 2 above, we have:

20, 20, 20, 20, 20, 20, 21, 21, 21, 21, 22, 22, 22, 22, 23, 23, 24, 24, 24, 25

Thus,

- 25th percentile (P_{25}) or 1st quartile (Q_1): as $i = \frac{20*25}{100} = \frac{500}{100} = 5$ is integer,

$$P_{25} = Q_1 = \frac{X_5 + X_6}{2} = \frac{20 + 20}{2} = 20$$

- 50th percentile (P_{50}) or median: as $i = \frac{20*50}{100} = \frac{1000}{100} = 10$ is integer,

$$P_{50} = Q_2 = \tilde{x} = \frac{X_{10} + X_{11}}{2} = \frac{21 + 22}{2} = 21.5$$

- 75th percentile (P_{75}) or 3rd quartile (Q_3): as $i = \frac{20*75}{100} = \frac{1500}{100} = 15$ is integer,

$$P_{75} = Q_3 = \frac{X_{15} + X_{16}}{2} = \frac{23 + 23}{2} = 23$$

Range

The range for a data set is the difference between the maximum value (greatest value) and the minimum value (lowest value) in the data set; that is:

$$range = maximum\ value - minimum\ value$$

The range should have the same units as those of the data values from which it is computed.

The interquartile range (IQR) is the difference between the first and third quartiles; that is, $IQR = Q_3 - Q_1$ (McCune, 2010).

In example 2 above, minimum value=20, maximum value=25. Thus, the range is given by 25-20=5.

Standard Deviation and Variance

The variance and standard deviation are widely used measures of variability. They provide a measure of the variability of a variable. It measures the offset from the mean of a variable. If there is no variability in a variable, each data value equals the mean, so both the variance and standard deviation for the variable are zero. The greater the distance of the variable' values from the mean, the greater is its variance and standard deviation.

The relationship between the variance and standard deviation measures is quite simple. The standard deviation (denoted by σ for population standard deviation and s for sample standard deviation) is the square root of the variance (denoted by σ^2 for population variance and s^2 for sample variance).

The formulas for variance and standard deviation (for population and sample, respectively) are:

- **Population Variance:** $\sigma^2 = \frac{\Sigma\left(x_i - \mu\right)^2}{N}$, where x_i is the ith data value from the population, μ is mean of the population, and N is the size of the population.
- **Sample Variance:** $s^2 = \frac{\Sigma\left(x_i - \overline{x}\right)^2}{n-1}$, where x_i is the ith data value from the sample, $\overline{x}$ is mean of the sample and n is the size of the sample.
- **Population Standard Deviation:** $\sigma = \sqrt{\sigma^2} = \sqrt{\frac{\Sigma\left(x_i - \mu\right)^2}{N}}$.
- **Sample Standard Deviation:** $s = \sqrt{s^2} = \sqrt{\frac{\Sigma\left(x_i - \overline{x}\right)^2}{n-1}}$.

Charts and Graphs

Data can be summarized in a visual way using charts and/or graphs. These are displays that are organized to give a big picture of the data in a flash and to zoom in on a particular result that was found. Depending on the data type, the graphs include pie charts, bar charts, time charts, histograms or boxplots.

Pie Charts

A pie chart (or a circle chart) is a circular graphic. Each category is represented by a slice of the pie. The area of the slice is proportional to the percentage of responses in the category. The sum of all slices of the pie should be 100% or close to it (with a bit of round-off error). The pie chart is used with categorical variables or discrete numerical variables. Figure 1 represents the example 1 above.

Bar Charts

A bar chart (or bar graph) is a chart that presents grouped data with rectangular bars with lengths proportional to the values that they represent. The bars can be plotted vertically or horizontally. A vertical bar chart is sometimes called a column bar chart. In general, the x-axis represents categorical variables or discrete numerical variables. Figure 2 and Figure 3 represent the example 1 above.

Figure 1. Pie chart example

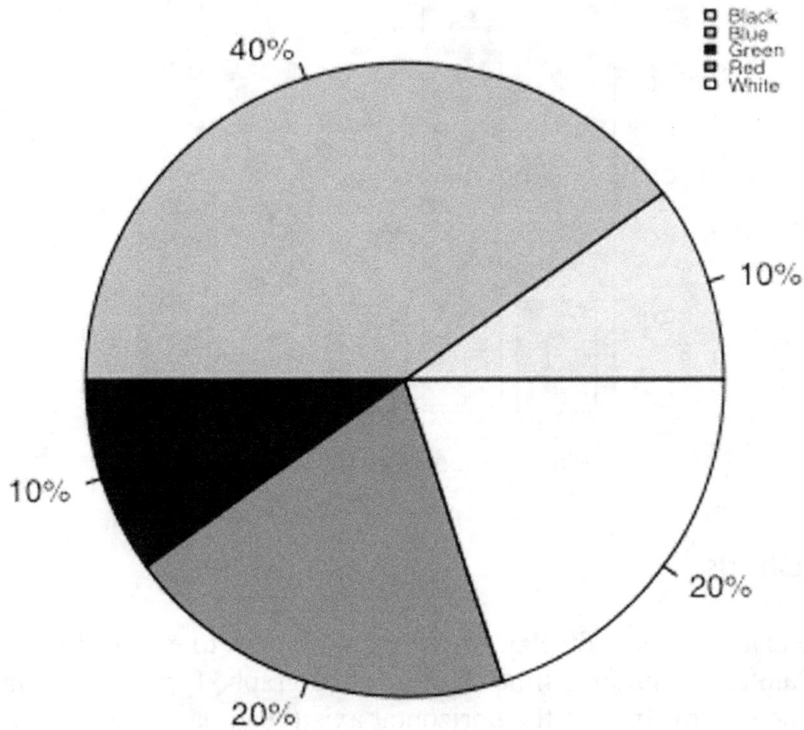

Figure 2. Bar graph example (with frequencies)

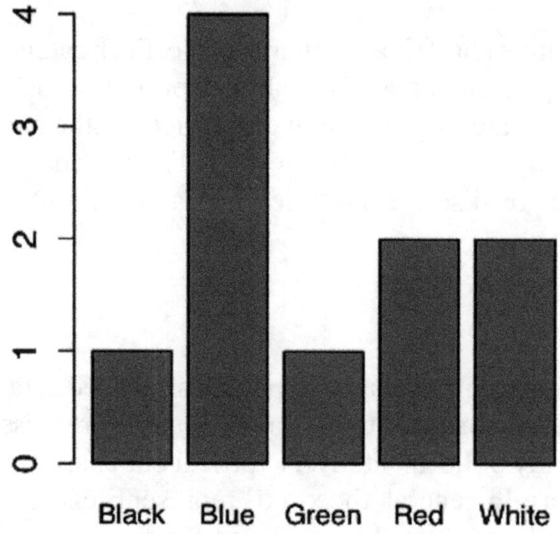

Figure 3. Bar graph example (with relative frequencies)

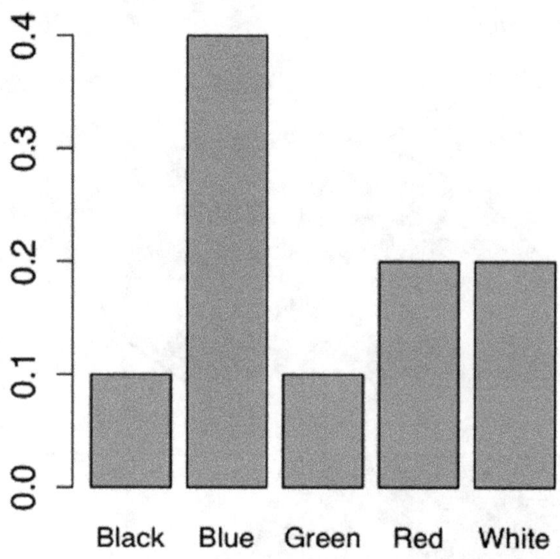

Time Charts

A time chart is a data display whose main point is to examine trends over time. Another name for a time chart is a line graph. Typically, a time chart has some unit of time on the horizontal axis (year, day, month, and so on) and a measured quantity on the vertical axis (average household income, birth rate, total sales, or others). At each time's period, the amount is shown as a dot, and the dots are connected to form the time chart (Rumsey, 2010).

Figure 4 is an example of a time chart. It represents the number of accidents, for instance, in a small city along some years.

Histogram

A histogram is a graphical representation of numerical data distribution. It is an estimate of the probability distribution of a continuous quantitative variable. Because the data is numerical, the categories are ordered from smallest to largest (as opposed to categorical data, such as gender, which has no inherent order to it). To be sure each number falls into exactly one group, the bars on a histogram touch each other but don't overlap (Rumsey, 2010). The height of a bar in a histogram may represent either frequency or a percentage (Peers, 2006). Figure 5 accounts for the histogram of example 3 above.

Boxplot

A boxplot or box plot is a convenient way of graphically depicting groups of numerical data. It is a one-dimensional graph of numerical data based on the five-number summary, which includes the minimum value, the 25th percentile (also known as Q_1), the median, the 75th percentile (Q_3), and the

Figure 4. Time chart example

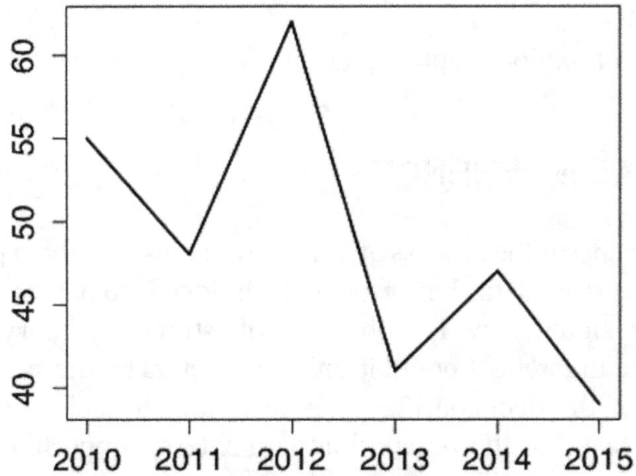

Figure 5. Histogram example

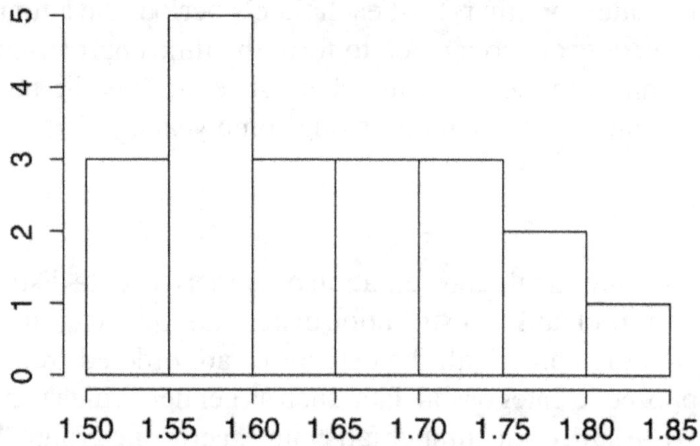

maximum value. These five descriptive statistics divide the data set into four equal parts (Rumsey, 2010).

Some statistical software adds asterisk signs (*) or circle signs (o) to show numbers in the data set that are considered to be, respectively, outliers or suspected outliers — numbers determined to be far enough away from the rest of the data. There are two types of outliers:

1. **Outliers:** Either $3 \times IQR$ or more above the third quartile or $3 \times IQR$ or more below the first quartile.
2. **Suspected Outliers:** Slightly more central versions of outliers: either 1.5×IQR or more above the third quartile or 1.5×IQR or more below the first quartile.

Figure 6 is a boxplot's representation.

STATISTICAL INFERENCE

Statistical inference is the process of drawing conclusions about populations or scientific truths from data. This process is divided into two areas: estimation theory and decision theory. The objective of estimation theory is to estimate the value of the theoretical population's parameters by the sample forecasts. The purpose of the decision theory is to establish decisions with the use of hypothesis tests for the population parameters, supported by a concrete

Figure 6. Boxplot

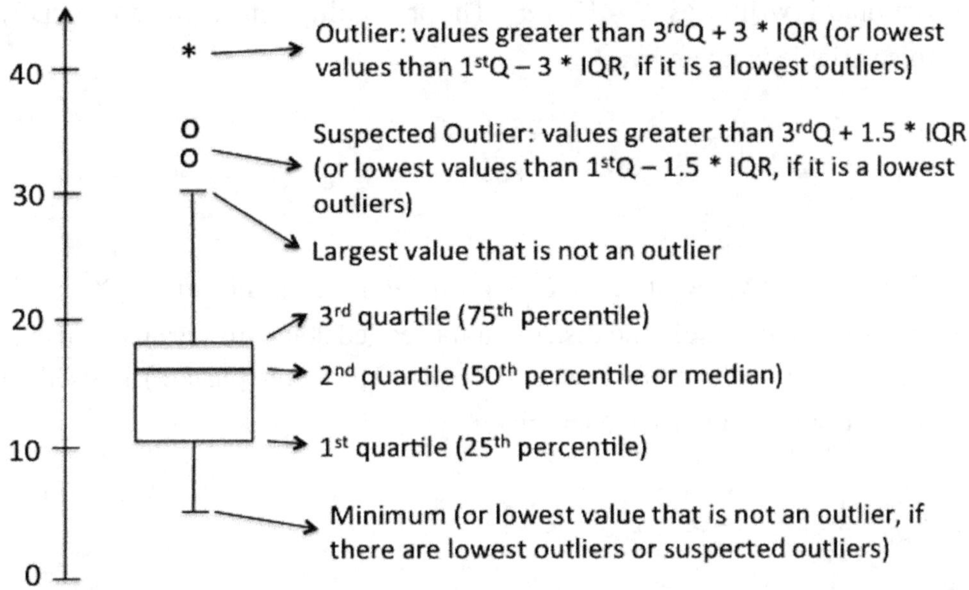

measure of the degree of certainty/uncertainty regarding the decision that was taken (Marôco, 2011).

Inference Distribution Functions (Most Frequent)

The statistical inference process requires that the probability density function (a function that gives the probability of each observation in the sample) is known, that is, the sample distribution can be estimated. Thus, the common procedure in statistical analysis is to test whether the observations of the sample are properly fitted by a theoretical distribution. Several statistical tests (e.g., the Kolmogorov-Smirnov test or the Shapiro-Wilk test) can be used to check the sample adjustment distributions for particular theoretical distribution. The following distributions are some probability density functions commonly used in statistical analysis.

Normal Distribution

The normal distribution or Gaussian distribution is the most important probability density function on statistical inference. The requirement that the sampling distribution is normal is one of the demands of some statistical methodologies with frequent use, called parametric methods (Marôco, 2011).

A random variable X with a normal distribution of mean μ and standard deviation σ is written as $X \sim N(\mu, \sigma)$. The probability density function (PDF) of this variable is given by:

$$f_X(x) = \frac{1}{\sigma\sqrt{2\pi}} e^{-\frac{1}{2}\left(\frac{x-\mu}{\sigma}\right)^2}, -\infty \le x \le +\infty$$

The expected value of X is $E(X) = \mu$, and the variance is $V(X) = \sigma^2$. When $\mu = 0$ and $\sigma = 1$, the distribution is called standard normal distribution and is typically written as $Z \sim N(0,1)$. The letter phi (φ) is used to denote the standard normal PDF given by:

$$\varphi(z) = \frac{1}{\sqrt{2\pi}} e^{-\frac{1}{2}(z^2)}, -\infty \le z \le +\infty$$

The normal distribution graph has a bell-shaped line (one of the normal distribution names is bell curve) and is completely determined by the mean and standard deviation of the sample. Figure 7 shows a distribution $N(0,1)$. See also Table 7.

Figure 7. Normal distribution

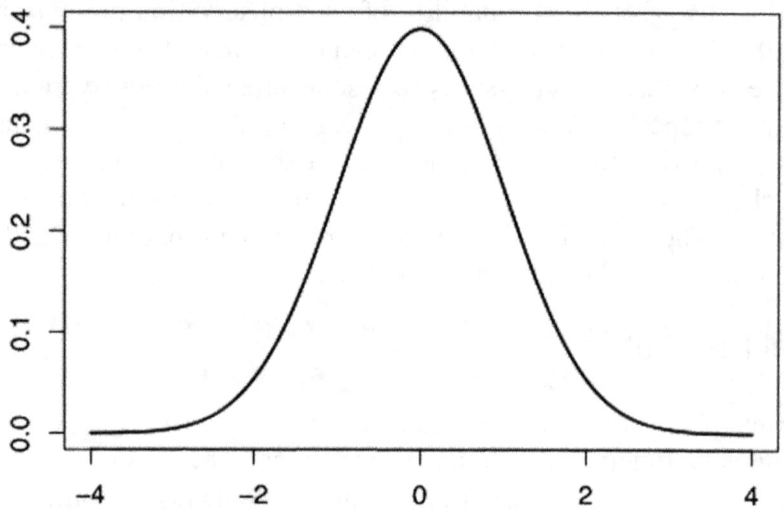

Table 7. Normal distribution and standard deviation intervals

Range	Proportion
$\mu \pm 1\sigma$	68.3%
$\mu \pm 2\sigma$	95.5%
$\mu \pm 3\sigma$	99.7%

Although there are many normal curves, they all share an important property that allows us to treat them in a uniform fashion. Thus, all normal density curves satisfy the property shown in Table 7, which is often referred to as the Empirical Rule. Thus, for a normal distribution, almost all values lie within three standard deviations of the mean.

Chi-Square Distribution

A random variable X obtained by the sums of squares of n random variables $Z_i \sim N(0,1)$ has a chi-square distribution with n degrees of freedom, denoted as $X^2(n)$. The probability density function (PDF) of this variable is given by (Kerns, 2010):

$$f_X(x) = \frac{1}{2^{\frac{n}{2}} \cdot \int_0^{+\infty} x^{\frac{n}{2}-1} \cdot e^{-X} \cdot dX} \cdot x^{\frac{n}{2}-1} \cdot e^{-\frac{x}{2}}$$

with $n > 0$ e $x > 0$. Figure 8 shows an example of a chi-square distribution.

The expected value of X is $E(X) = n$ and the variance is $V(X) = 2n$. As noted above, the X^2 distribution is the sum of squares of n variables $N(0,1)$. Thus, the central limit theorem (see section *central limit theorem*) also ensures that the X^2 distribution approaches the normal distribution for high values of p.

Figure 8. Chi-square distribution example

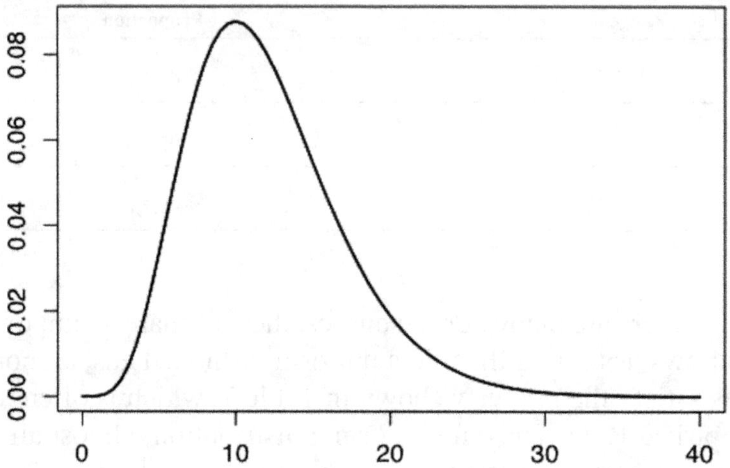

Student's t-Distribution

Student's t-distribution is a probability distribution that is used to estimate population parameters when the sample size is small and/or when the population variance is unknown.

A random variable $X = \dfrac{Z}{\sqrt{Y / n}}$ has a student's t-distribution with n degrees of freedom, if $Z \sim N(0,1)$, and $Y \sim X^2(n)$ are independent variables. The probability density function (PDF) of this variable is given by (Kerns, 2010):

$$f_X(x) = \frac{\tau\left(\frac{n+1}{2}\right)}{\sqrt{n\pi} \cdot \tau\left(\frac{n}{2}\right)} \cdot \left(1 + \frac{x^2}{n}\right)^{-\frac{1}{2}(n+1)}, -\infty < x < +\infty$$

where

$$\tau(u) = \int_0^{+\infty} x^{u-1} \cdot e^{-x} \cdot dX$$

and $n > 0$. When n increases, this distribution approximates to the centered reduced normal distribution ($N(0,1)$). Figure 9 shows an example of a student's t-distribution.

As the centered reduced normal distribution, the student's t-distribution has expected value $E(X) = 0$ and variance $V(X) = \frac{n}{n-2}, n > 2$.

Snedecor's F-Distribution

Snedecor's F-distribution is a continuous statistical distribution which arises in the testing of whether two observed samples have the same variance.

A random variable $X = \frac{\frac{Y1}{m}}{\frac{Y2}{n}}$ where $Y_1 \sim X^2(m)$ and $Y_2 \sim X^2(n)$, has a Snedecor's F-distribution with m and n degrees of freedom, $X \sim F(m,n)$. The probability density function (PDF) of this variable is given by (Kerns, 2010):

$$f_X(x) = \frac{\tau\left(\frac{m+n}{2}\right)}{\tau\left(\frac{m}{2}\right)\cdot\tau\left(\frac{n}{2}\right)}\cdot\left(\frac{m}{n}\right)^{\frac{m}{2}}\cdot x^{\frac{m}{2}-1}\cdot\left(1+\frac{m}{n}x\right)^{-\frac{m+n}{2}}, x > 0$$

Figure 9. Student's t-distribution example

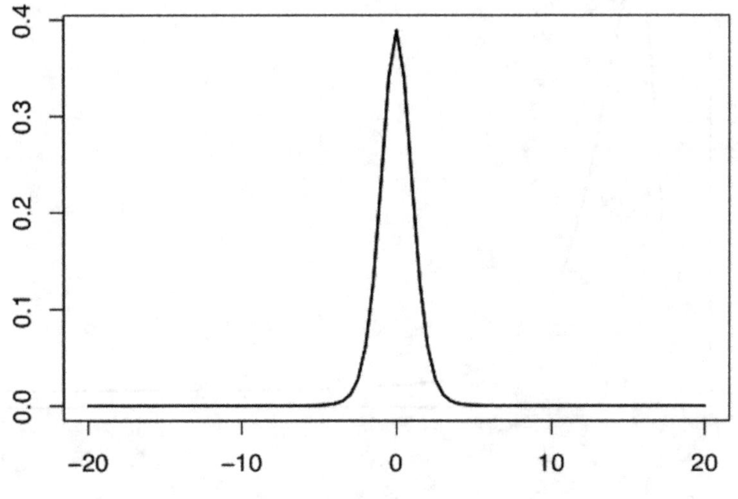

where

$$\tau(u) = \int_{0}^{+\infty} x^{u-1} \cdot e^{-x} \cdot dX$$

and $m > 2$ and $n > 4$. Figure 10 shows an example of a Snedecor's F-distribution.

The expected value of X is $E(X) = \frac{n}{n-2}$ with $n > 2$ and the variance is:

$$V(X) = \frac{2n^2 \cdot (m+n-2)}{m \cdot (n-2)^2 \cdot (n-4)}.$$

Binomial Distribution

The binomial distribution is the discrete distribution most used in statistical inference to test hypotheses concerning proportions of dichotomous nominal variables (true vs. false, exist vs. non-exists). This distribution is obtained with exactly n successes out of N Bernoulli trials (where the result of each Bernoulli trial is true with probability p and false with probability $q = 1 - p$).

Figure 10. Snedecor's F-distribution example

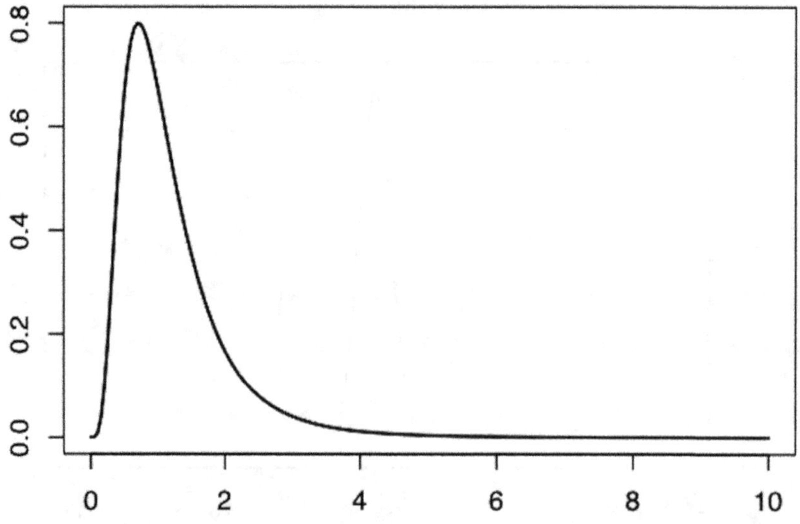

The binomial distribution for the variable X has n and p parameters and is denoted as $X \sim B(n,p)$. The probability mass function (PMF) of this variable is given by:

$$f_X(x) = \binom{n}{x} p^x (1-p)^{n-x}, x = 0,1,2,\ldots,n$$

Figure 11 shows an example of a binomial distribution.

The expected value of variable X is $E(X) = n \cdot p$, and the variance is $V(X) = n \cdot p \cdot q$. Such as the chi-square distribution or student's t-distribution, the central limit theorem ensures that the binomial distribution is approximated by the normal distribution, when n and p are sufficiently large ($n > 20$ and $np > 7$; Marôco, 2011).

Sampling Distribution

To perform statistical inference - confidence intervals estimation or performing hypothesis testing – it is necessary to know the distributional properties of the sample, from which it is intended to infer for the theoretical population (Marôco, 2011). In the examples given so far, a population was specified, and the sampling distribution of the mean and the range were determined. In

Figure 11. Binomial distribution example

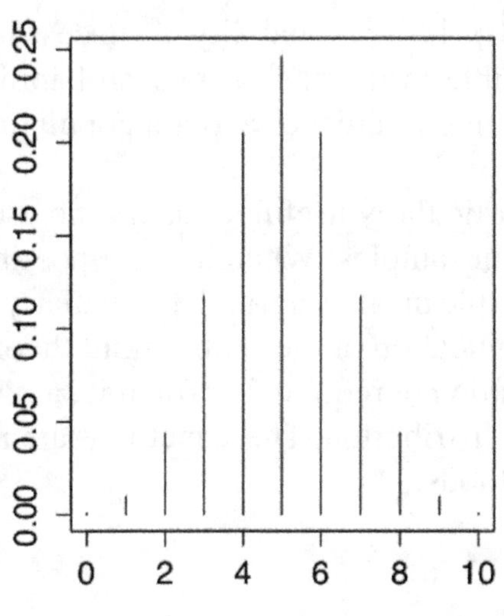

practice, the process proceeds the other way: the sample data is collected, and from these data, the parameters of the sampling distribution are estimated. The mean of a representative sample provides an estimate of the unknown population mean, but intuitively we know that if we took multiple samples from the same population, the estimates would vary from one another. We could, in fact, sample over and over from the same population and compute a mean for each of the samples. All these sample means constitute yet another "population", and we could graphically display the frequency distribution of the sample means. This is referred to as the sampling distribution of the sample means.

Some of the sampling distributions commonly used in statistical inference process are presented in the Table 8 (Marôco, 2011).

The sample's mean is one of the most relevant statistics for both the theory of estimation as to the theory of decision.

CENTRAL LIMIT THEOREM

The central limit theorem claims that the distribution of the sample means will be approximately normally distributed if the population has mean μ and standard deviation σ, and take sufficiently large random samples from the population with replacement. This will hold true regardless of whether the source population is normal or skewed, provided the sample size is sufficiently large (usually $n > 30$). If the population is normal, then the theorem holds true even for samples smaller than 30. In fact, this also holds true even if the population is binomial, provided that $\min\left(np, n\left(1-p\right)\right) > 5$, where n is the sample size and p is the probability of success in the population. This means that it is possible to use the normal probability model to quantify uncertainty when making inferences about a population mean based on the sample mean.

This theorem is particularly useful to justify the use of parametric methods for high dimension samples. When it is not possible to assume that the distribution of the sample mean is normal, particularly when the sample size does not allow the application of the central limit theorem, it is necessary to resort to methods that do not require, in principle, any assumption about the form of the sampling distribution. These methods are referred to generically as nonparametric methods.

Table 8. Sampling distributions commonly used in statistical inference

Statistic	Sampling Distribution
$\bar{X}$	$\bar{X} \sim N\left(\mu, \frac{\sigma}{\sqrt{n}}\right)$ if the sampling is with replacement or if the population is too large. $\bar{X} \sim N\left(\mu, \frac{\sigma}{\sqrt{n}} \times \frac{N-n}{N-1}\right)$ if the sampling is without replacement or if the population is small $\left(\frac{n}{N}\right) \leq 0.05$. $\frac{\bar{X}-\mu}{\frac{S'}{\sqrt{n}}} \sim t\left(n-1\right)$ if the population standard deviation is unknown.
S'^2	$\frac{\left(N-1\right)S'^2}{\sigma^2} \sim X^2_{\left(n-1\right)}$ if the variable has normal distribution
$\frac{S'^2_A}{S'^2_B}$	$\frac{S'^2_A}{S'^2_B} \sim F\left(n_A-1, n_B-1\right)$ if the variances have X^2 distribution
$\hat{P}$	$\hat{P} \sim B\left(n,p\right)$ for small samples $\frac{\hat{p}-p}{\sqrt{\frac{\hat{p}\left(1-\hat{p}\right)}{n}}} \sim N\left(0,1\right)$ for large samples (with $n > 20$ e $np > 5$, where p is the population proportion)

Marôco, 2011.

HYPOTHESIS TESTS

A statistical hypothesis is an assumption about a population parameter. This assumption may or may not be true. Hypothesis tests refer to the formal procedures used by statisticians to accept or reject a statistical hypothesis.

The best way to determine whether a statistical hypothesis is true would be to examine the entire population. Since that is often impractical, statistical tests are used to determine whether there is enough evidence in a sample of data to infer that a particular condition is true for the entire population. If sample data are not consistent with the statistical hypothesis, the hypothesis is rejected.

Hypothesis tests examine two opposing hypotheses about a population: the null hypothesis and the alternative hypothesis.

The null hypothesis, denoted by H_0, is the statement being tested. Usually, the null hypothesis is a

declaration of the absence of effect or no effect at all and less compromising. The alternative hypothesis, denoted by H_1, is the hypothesis that sample observations are influenced by some non-random cause.

The H_0 should only be rejected if there is enough evidence for a given probability of error or a certain level of confidence, which suggests in fact H_0 is not valid.

However, a hypothesis test can have one of two outcomes: the reader accepts the null hypothesis, or it rejects the null hypothesis. Many statisticians stress with the notion of "accepting the null hypothesis". Instead, they say: you reject the null hypothesis, or you fail to reject the null hypothesis. The distinction between "acceptance" and "failure to reject" is crucial. Whilst acceptance implies that the null hypothesis is true, failure to reject means that the data is not sufficiently persuasive to prefer the alternative hypothesis to the null hypothesis.

A hypothesis test is developed in the following steps:

- **State the Hypotheses:** This involves stating the null and alternative hypotheses. The hypotheses are stated in such a way that they are mutually exclusive. That is, if one is true, the other must be false.
- **Formulate an Analysis Plan:** The analysis plan describes how to use sample data to evaluate the null hypothesis. The evaluation often focuses around a single test statistic.
- **Analyze Sample Data:** Find the value of the test statistic (mean score, proportion, t-score, z-score, etc.) described in the analysis plan.
- **Interpret Results:** Apply the decision rule described in the analysis plan. If the value of the test statistic is unlikely, based on the null hypothesis, reject the null hypothesis.

When considering whether the null hypothesis is rejected and the alternative hypothesis is accepted, it is needed to find the direction of the alternative hypothesis statement. This could be a one-tailed test or two-tailed test.

A one-tailed test is a statistical test in which the critical area of the distribution is one-sided so that it is either greater than or less than a particular value, but not both. If the sample that is being tested falls into the one-sided critical area, the alternative hypothesis will be accepted instead of the null hypothesis. The one-tailed test gets its name from checking the area under one of the tails (sides) of a normal distribution, although the test can be used in other non-normal distributions as well.

For example, suppose the null hypothesis states that the mean is less than or equal to 10. The alternative hypothesis would be that the mean is greater than 10. The region of rejection would consist of a range of numbers located on the right side of sampling distribution; that is, a set of numbers greater than 10. This represents the implementation of a one-tailed test.

A two-tailed test is a statistical test in which the critical area of the distribution is two sided and tests whether a sample is either greater than or less than a specified range of values. If the sample that is being tested falls into either of the critical areas, the alternative hypothesis will be accepted instead of the null hypothesis. The two-tailed test gets its name from checking the area under both of the tails (sides) of a normal distribution, although the test can be used in other non-normal distributions.

For example, suppose the null hypothesis states that the mean is equal to 10. The alternative hypothesis would be that the mean is different to 10, i.e., less than 10 or greater than 10. The region of rejection would consist of a range of numbers located on both sides of sampling distribution; that is, the region of rejection would consist partly of numbers that were less than 10 and partly of numbers that were greater than 10.

DECISION RULES

The analysis plan includes decision rules for rejecting the null hypothesis. In practice, statisticians describe these decision rules in two ways - concerning a *p*-value or concerning a region of acceptance.

p-Value and Statistical Errors

The *p*-value is the probability of observing a value of the test statistic as extreme or more extreme than the observed test statistic that you computed from the sample. Regarding the distribution associated with the hypothesis test, the *p*-value is calculated as follows:

- For a one-tailed test, the *p*-value is the area to the right (right-tailed test) or left (left-tailed test) of the test statistic.

- For a two-tailed test, the *p*-value is two times the area to the right of a positive test statistic or the left of a negative test statistic.

To make a decision about rejecting or not rejecting H_0, it is necessary to determine the cutoff probability for the *p*-value before doing a hypothesis test; this cutoff is called an alpha level (α). Typical values for α are 0.05 or 0.01.

When *p*-value (instead of the test statistic) is used in the decision rule, the rule becomes: If the *p*-value is less than α (the level of significance), reject H_0 and accept H_1. Otherwise, fail to reject H_0.

However, incorrect interpretations of *p*-values are very common. The most common mistake is to interpret a *p*-value as the probability of making an error by rejecting a true null hypothesis (called a type I error).

There are several reasons why *p*-values can't be the error rate.

First, *p*-values are calculated based on the assumptions that the null is true for the population and that the difference in the sample is caused entirely by random chance. Consequently, *p*-values can't tell the probability that the null hypothesis is true or false because it is 100% true from the perspective of the calculations.

Second, while a small *p*-value indicates that the data are unlikely assuming a true null, it can't evaluate which of two competing cases is more likely: 1) The null is true, but the sample was unusual or; 2) The null is false. Determining which case is more likely requires subject area knowledge and replicate studies.

For example, supposing that a vaccine study produced a *p*-value of 0.04. The correct way to interpret this value is: assuming that the vaccine had no effect, it would obtain the observed difference or more in 4% of studies due to random sampling error. An incorrect way to interpret is: if the null hypothesis is rejected there is a 4% chance that a mistake is being made.

Types of Errors

The point of a hypothesis test is to make the correct decision about H0. Unfortunately, hypothesis testing is not a simple matter of being right or wrong. No hypothesis test is 100% certain because the hypothesis test is based on probability, so there is always a chance that an error has been made. Two types of errors are possible: type I and type II. The risks of these two errors are inversely related and determined by the significance level and the power for the test.

Table 9 shows the four possible situations.

Table 9. Possible situations after hypothesis testing

		Decision	
		Fail to Reject	**Reject**
Null Hypothesis	**True**	Correct Decision (probability = 1 - α)	**Type I Error** - rejecting the null when it is true (probability = α)
	False	**Type II Error** - fail to reject the null when it is false (probability = β)	Correct Decision (probability = 1 - β)

Type I Error

When the null hypothesis is true, and it is rejected, it has a type I error. The probability of making a type I error is α, which is the significance level set for the hypothesis test. An α of 0.05 indicates that it is willing to accept a 5% chance that being wrong when rejecting the null hypothesis. To reduce this risk, a lower value for α should be used. However, using a lower value for alpha, it will be less likely to detect a true difference if one exists.

Type II Error

When the null hypothesis is false, and it is failed to reject it, it has a type II error. The probability of making a type II error is β, which depends on the power of the test. It is possible to decrease the risk of committing a type II error by providing that the test has enough power. Ensuring the sample size is large enough to detect a practical difference when one truly exists can do this.

The probability of rejecting the null hypothesis when it is false is equal to 1–β. This value is the power of the test.

The following example helps to understand the interrelationship between type I, and type II error, and to determine which error has more severe consequences for each situation. If there is interest in comparing the effectiveness of two medications, the null and alternative hypotheses are:

- **Null Hypothesis (H_0): μ1= μ2:** The two medications have equal effectiveness.
- **Alternative Hypothesis (H_1): μ1≠ μ2:** The two medications do not have equal effectiveness.

A type I error occurs if the null hypothesis is rejected, i.e., if it is possible to conclude that the two medications are different when, in fact, they are not. If the medications have the same effectiveness, this error may not be considered too severe because the patients still benefit from the same level of effectiveness regardless of which medicine they take.

However, if a type II error occurs, the null hypothesis is not rejected when it should be rejected. That is, it is possible to conclude that the medications have the same effectiveness when, in fact, they are different. This error is potentially life-threatening if the less-effective drug is sold to the public instead of the more effective one.

When the hypothesis tests are conducted, consider the risks of making type I and type II errors. If the consequences of making one type of error are more severe or costly than making the other type of error, then choose a level of significance and power for the test that will reflect the relative severity of those consequences.

Acceptance Region vs. Rejection Region

The acceptance region is a range of values. If the test statistic falls within the region of acceptance, the null hypothesis is not rejected. The acceptance region is defined so that the chance of making a type I error is equal to the significance level.

The set of values outside the acceptance region is called the rejection region. If the test statistic falls within the rejection region, the null hypothesis is rejected. The rejection region is also known as the critical region. The value(s) that separates the critical region from the acceptance region is called the critical value(s). In such cases, we say that the hypothesis has been rejected at the α level of significance.

Confidence Intervals

A confidence interval is an estimated range of a parameter of a population. Instead of estimating the parameter by a single value, it is given a range of probable estimates.

Confidence intervals are used to indicate the reliability of an estimate. For example, a confidence interval can be used to describe how the results of a search are trustworthy. If all the estimates are equals, a search that results in a small confidence interval is more reliable than one that results in a higher confidence interval. These intervals are usually calculated so that this percentage is 95%, but it can produce 90%, 99%, 99.9% (or whatever) confidence intervals for the unknown parameter.

The width of the confidence interval gives some idea of how uncertain the research is about the unknown parameter. A very wide interval may indicate that more data should be collected before anything very definite can be said about the parameter.

Confidence intervals are more informative than the simple results of hypothesis tests (where we decide "reject H_0" or "don't reject H_0") since they provide a range of plausible values for the unknown parameter.

Confidence limits are the lower and upper boundaries/values of a confidence interval, that is, the values that define the range of a confidence interval.

The upper and lower bounds of a 95% confidence interval are the 95% confidence limits. These limits may be taken for other confidence levels, for example, 90%, 99%, and 99.9%.

The confidence level is the probability value $1-\alpha$ associated with a confidence interval.

It is often expressed as a percentage. For example, say $\alpha = 0.05 = 5\%$, then the confidence level is equal to $1-0.05 = 0.95$, i.e. a 95% confidence level. For example, suppose an opinion poll predicted that, if the election were held today, the Conservative party would win 60% of the vote. The pollster might attach a 95% confidence level to the interval 60% plus or minus 3%. That is, he thinks it very likely that the Conservative party would get between 57% and 63% of the total vote.

Summarizing:

- A *p*-value is a probability of obtaining an effect as large as or greater than the observed effect, assuming null hypothesis is true.
  - Provides a measure of strength of evidence against the H_0.
  - Does not provide information on magnitude of the effect.
  - Affected by sample size and magnitude of effect: interpret with caution!
  - Cannot be used in isolation to inform clinical judgment.
- Confidence interval quantifies:
  - How confident are we about the true value in the source population.
  - Better precision with large sample size.
  - Corresponds to hypothesis testing, but much more informative than *p*-value.
- Keep in mind clinical importance when interpreting statistical significance!

Parametric and Non-Parametric Tests

During the process of statistical inference, there is often the question about the best hypothesis test for data analysis. In statistics, the test with higher power $(1-\beta)$ is considered the most appropriate and more robust to violations of assumptions or application conditions.

Hypothesis tests are categorized into two major groups: parametric tests and non-parametric tests.

Parametric tests use more information than non-parametric tests and are, therefore, more powerful. However, if a parametric test is wrongly used with data that doesn't satisfy the needed assumptions, it may determine significant differences when truly there isn't one.

Alternatively, non-parametric tests use less information and, therefore, are more conservative tests than their parametric alternatives. This means that if the reader uses a non-parametric test when he/she has data that satisfies assumptions for a parametric test, the reader can decrease his/her power (i.e. he/she is less likely to get a significant result when, in reality, one exists: significant relationship, significant difference, or other).

CONCLUSION

This chapter presents the main concepts used in statistical analysis. Without these, it will be difficult for the reader to understand additional analysis that will be held in the course of this book.

The reader should now be able to recognize the used concepts, their meaning and when they should be applied.

The theoretical concepts presented in this chapter are:

- Variable, population and sample.
- Mean, median, mode, standard deviation, quartile and percentile.
- Statistic distributions:
  - Normal distribution.
  - Chi-square distribution.
  - Student's t-distribution.
  - Snedecor's F-distribution.
  - Binomial distribution.
- Central limit theorem.
- Decision rules: *p*-value, error, confidence interval and tests.

REFERENCES

Kerns, G. J. (2010). Introduction to probability and statistics using r. Lulu.com.

Marôco, J. (2011). Análise Estatística com o SPSS Statistics (5th ed.). Pero Pinheiro.

McCune, S. (2010). *Practice Makes Perfect Statistics* (1st ed.). McGraw-Hill.

Peers, I. (2006). *Statistical analysis for education and psychology researchers: Tools for researchers in education and psychology*. Routledge.

Rumsey, D. (2010). *Statistics Essentials for Dummies*. Wiley Publishing, Inc.

Chapter 2
Introduction to Programming R and Python Languages

INTRODUCTION

This chapter introduces the basic concepts of using the languages we propose to approach the data analysis tasks. Thus, we first introduce some features of R and then we also present some necessary features of Python. We stress that we do not cover all features of both languages but the essential characteristics that the reader has to be aware of to progress in furthers stages of this book.

TOOLS

As previously stated, besides focusing on the statistical tasks later in this book, we will provide practice procedures and examples in both R and Python languages. There are many information sources about these languages. We will state a brief summary of both languages characteristics. We will start with R language. If the reader needs information for Python we make it available further in this chapter.

DOI: 10.4018/978-1-68318-016-6.ch002

R

R is a powerful programming language, used in statistical tasks, data analysis, numerical analysis and others. The main characteristics of R are:

- Powerful,
- Stable,
- Free,
- Programmable,
- Open Source Software,
- Directed to the Visualization of Data.

On the downside, R might not be initially suitable for everyone since it needs user inputs on the command line. We will deal with this in this chapter to make the reader's life easier.

First, the reader will need to install R for his/her operating system (OS). R is available for Mac, Windows, and Linux on a website. Figure 1 shows an overview of the website to download R.

Figure 1. Overview of the website to download R

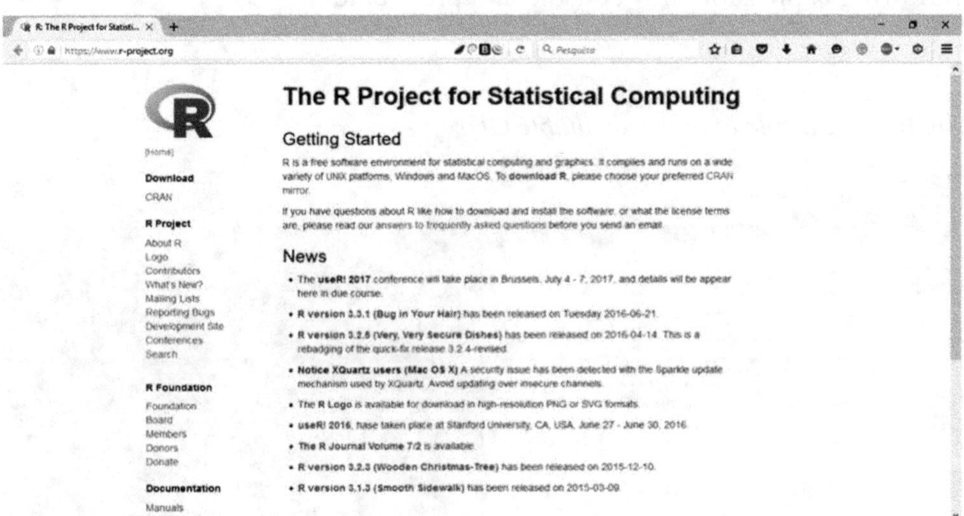

How to Use

R installation comes with a set of executables, including a GUI (Graphical User Interface) executable (for example, in windows, it is usually named as RGui.exe). Figure 2 shows this GUI.

In Figure 2, the reader has some information available, including the R version previously installed in the computer. Additionally, some commands are available to explore help regarding R commands. Figure 3 shows the use of the command *help.start()* and *q()*. The command *help.start()* opens a browser window with a manual browser version of R; this is represented in Figure 4. The *q()* command previously stated quits the R application.

In the manual browser, several links allow the reader to access documentation that can provide further assistance. The search for an answer in the manuals might be needed in the future adventure with R or any other language, so the reader should not be afraid to explore this documentation when needed.

A Session with R

With the RGui opened, the reader can try several commands, including expressions. For example, try to do a simple mathematics operation. Input the following expression in R console and press enter:

Figure 2. Example of an R executable GUI

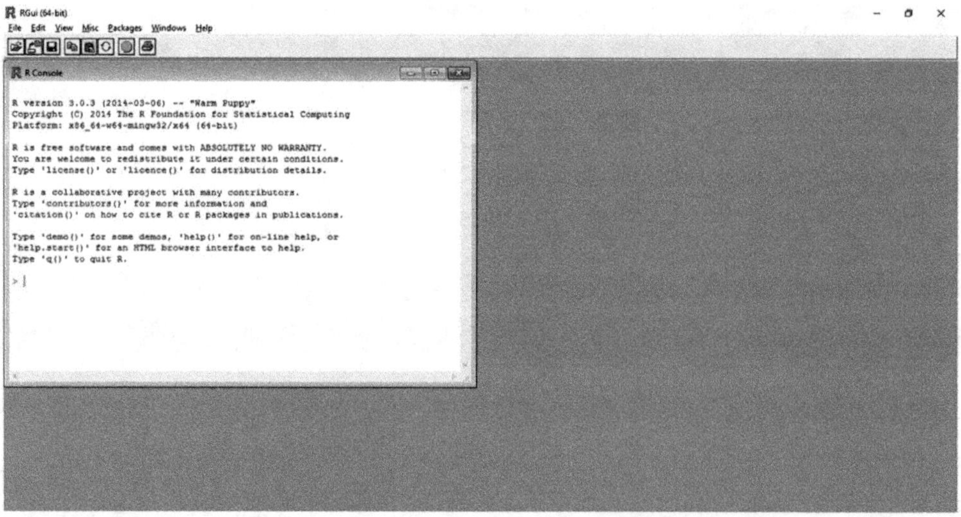

Figure 3. Example of some R commands

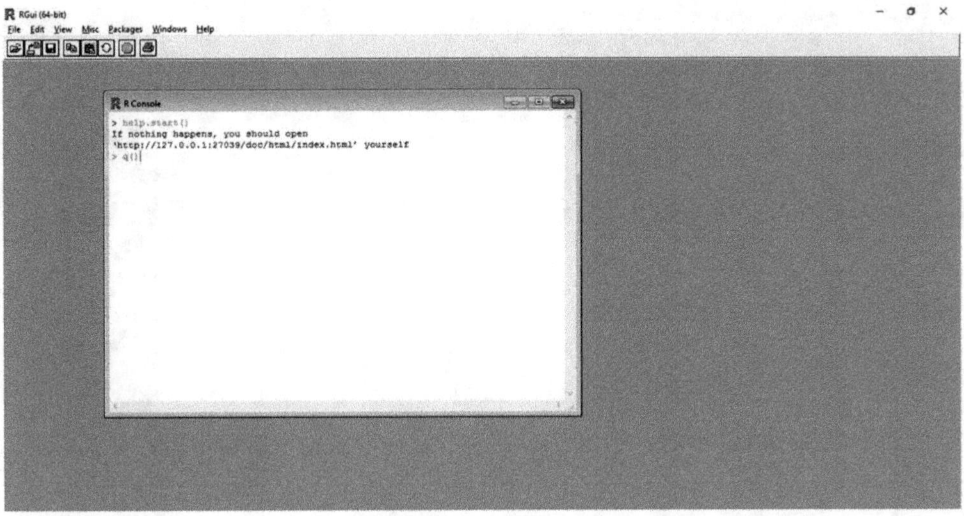

Figure 4. R manual (Internet browser version)

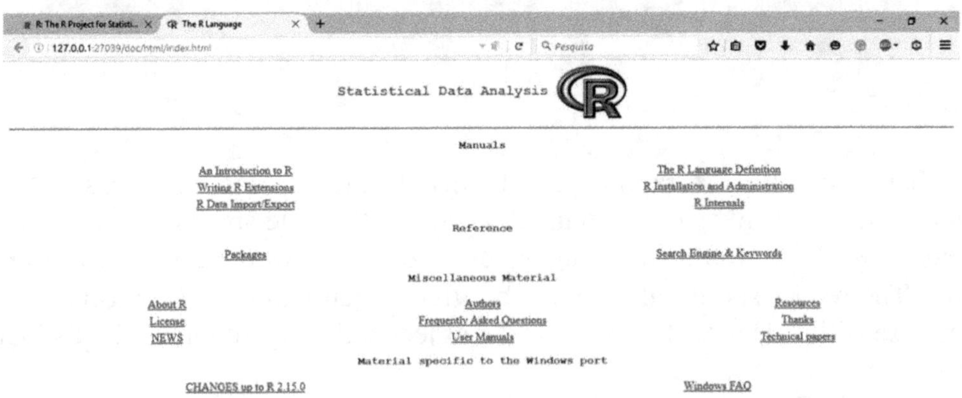

$3+2*5$

In Figure 5, the result provided in the R console is presented. It is clearly stated that the result is 13 in the line following the input command with the expression to solve.

Additionally, in Figure 5, we represent how to store the result of an expression in an object. In this example we stated that x equals three squared with the following expression:

Figure 5. Example of mathematics operations with R's console

$x \leftarrow 3\,\hat{}\,2$

The result, 9, is stored in the x object. The reader can name this object anything he/she likes unless named objects with white spaces. For example my_object_x would work nicely but *my object* x would give a syntax error. The reader should also remember that, when naming his objects, R is case sensitive. Hence, for example, an object called *X* is completely distinct to the object *x*.

Also, in Figure 1, we provide an insight of how to use the stored object *x*, this time, to obtain the square root of the value of *x* with the expression *sqrt(x)*. The reader probably has noticed that, when storing the result in *x*, the compiler did not provide a result for the expression in *x*. This can be done by just inputting a command with the object name *x* and pressing enter. Then, the compiler provides the result of the expression previously stored in *x*.

Installing RStudio

By now, the reader is probably asking himself/herself if there is a better way to work with R commands. It is clear that inputting a command one at each time and hitting enter, in the end, is too much of a workload. Thus, we will suggest the use of an Integrated Development Environment (IDE), to be able to work efficiently with R. There are several IDEs and GUIs available nowadays, like for example RCommander GUI and the RStudio IDE. We will proceed with the RStudio IDE. Figure 6 shows an overview of the site to download RStudio.

After installation, the reader should execute the RStudio program. Then, immediately notice four windows on the screen as appears in Figure 7. The upper left window is where the reader inserts the commands that wish R to run. In Figure 7 we entered the same previous code mentioned before when writing about R console commands.

Additionally, we can clearly see that in the lower left window, the R console also appears. This will be where the results of the commands appear. The upper right window shows the environment objects and again, the only object available at the moment; the *x* object is presented in this window as well as the value of the object after running the code we provided. The reader might be asking how to run the code by now. We have two choices, clicking the upper left window button named Run or the button Source. Nevertheless, these buttons have different behaviors. With the Run button, the code is executed one line at each time. Additionally, the parts of the code that were selected with the mouse can also be performed with the Run button. The reader could

Figure 6. Overview of the site to download RStudio IDE

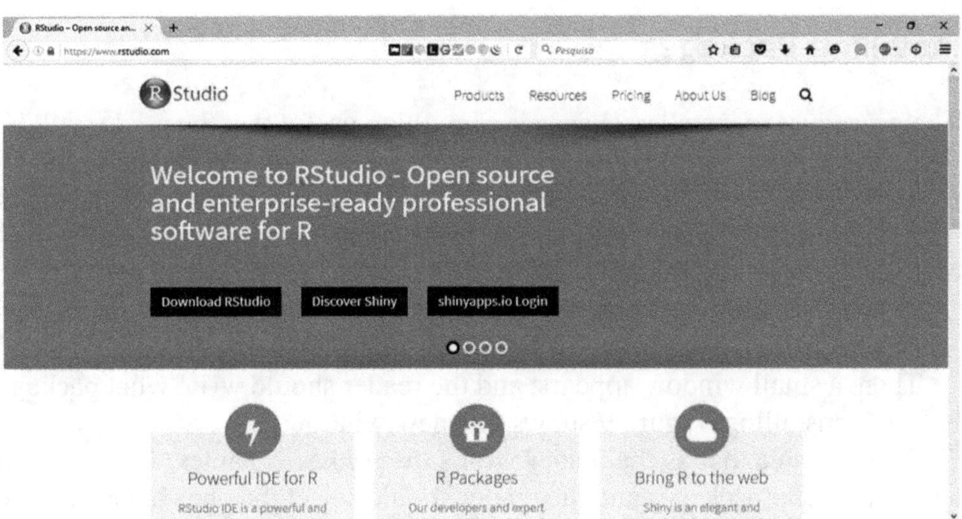

Figure 7. RStudio screen

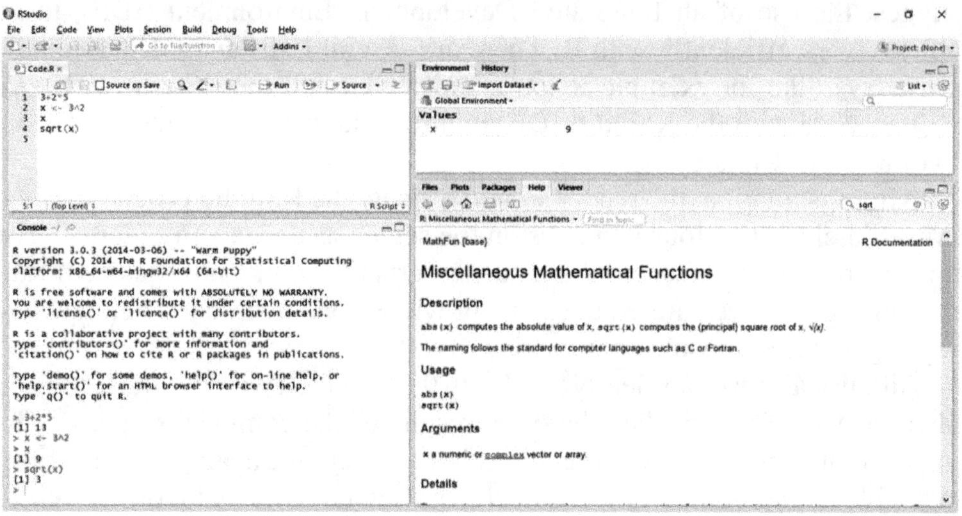

try, for example, to select only the line *sqrt(x)* and click Run. Only this line would be executed. By clicking Source instead, all the code present in the upper left window will run at once.

Finally, in the lower right window, several tabs will provide several types of experiments with R. Here, it is possible to have access to the R manual, and search by keywords through the guides. Additionally, this is also the window that will present plots or charts.

Installing Packages

The procedure to install packages is something useful that the reader will be doing throughout this book. Although R comes with many libraries already from the initial installation, there are many additional packages developed by the community. For certain tasks, these libraries are needed. Thus, it is required to install additional packages.

In RStudio, if the reader clicks on the Tools menu, one of the options the reader has is to install packages. Figure 8 represents these actions.

Then, a small window appears, and the reader should write what package he/she is installing. Figure 9 shows this new window.

Please notice that as the reader writes the packages names, RStudio will suggest several packages and the reader should select the ones he/she needs.

Figure 8. Tools to install packages in RStudio

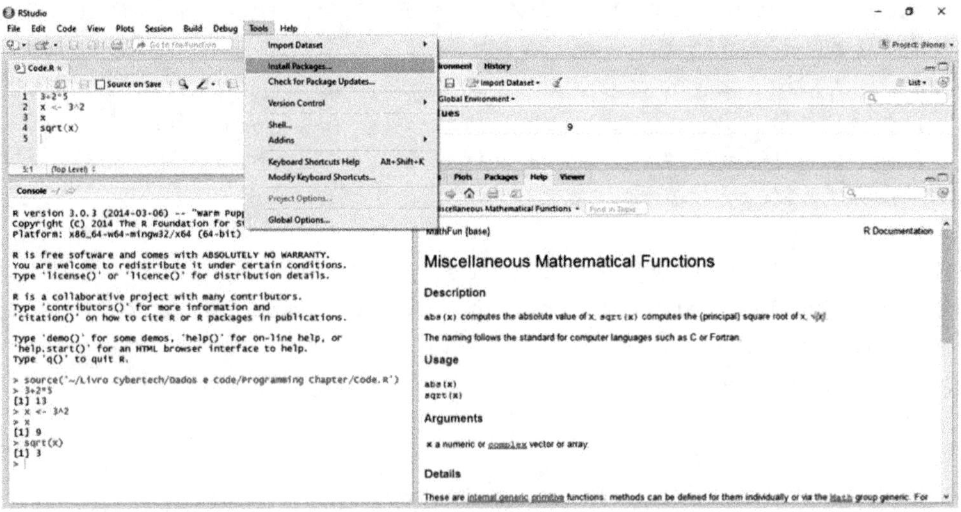

Figure 9. Install packages window (RStudio)

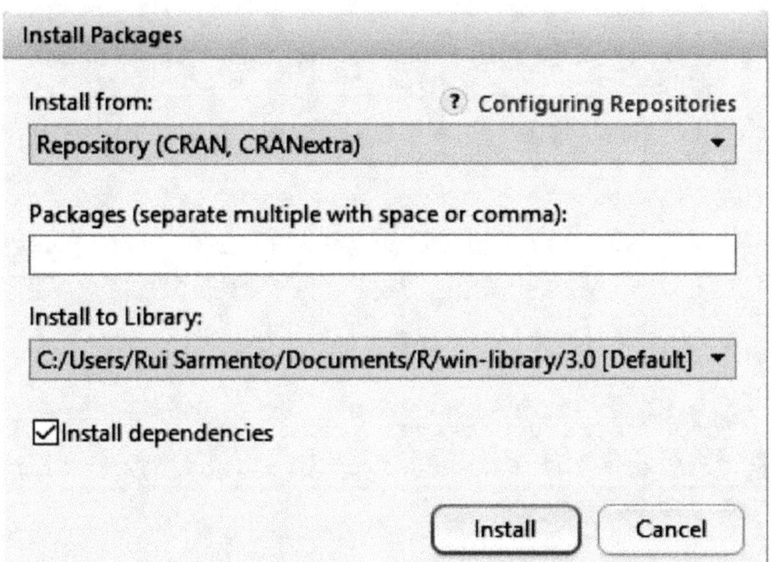

We present an example where we installed the package "StatRank". On the console, in the lower left window, what appears now is a description of the status of the package installation:

```
> install.packages("StatRank")
```

```
Installing package into 'C:/Users/Rui Sarmento/Documents/R/
win-library/3.0'
(as 'lib' is unspecified)
  There is a binary version available (and will be
  installed) but the source version is later:
         binary source
StatRank  0.0.4  0.0.6
also installing the dependencies 'evd', 'truncdist'
trying URL 'https://cran.rstudio.com/bin/windows/con-
trib/3.0/evd_2.3-0.zip'
Content type 'application/zip' length 1176785 bytes (1.1 Mb)
opened URL
downloaded 1.1 Mb
trying URL 'https://cran.rstudio.com/bin/windows/con-
trib/3.0/truncdist_1.0-1.zip'
Content type 'application/zip' length 26454 bytes (25 Kb)
opened URL
downloaded 25 Kb
trying URL 'https://cran.rstudio.com/bin/windows/con-
trib/3.0/StatRank_0.0.4.zip'
Content type 'application/zip' length 147850 bytes (144 Kb)
opened URL
downloaded 144 Kb
package 'evd' successfully unpacked and MD5 sums checked
package 'truncdist' successfully unpacked and MD5 sums
checked
package 'StatRank' successfully unpacked and MD5 sums
checked
The downloaded binary packages are in
        C:\Users\Rui Sarmento\AppData\Local\Temp\RtmpgB-
Jthk\downloaded_packages
```

As we have selected the option to install dependencies (recommended), RStudio has proceeded with the download of the needed files from the Internet and installed all required packages including packages dependencies, which are other packages themselves.

Vectors

Vectors are a typical structure in programming. The reader can store several values of the same type in a vector. Imagine a train composition with several coaches. Each wagon would be a position of the vector (the train), and each coach has a stored value. For example, we represent a vector of integer values from 1 to 10 this way:

```
> vector <- 1:10
> vector
 [1]  1  2  3  4  5  6  7  8  9 10
```

If the reader needs to do an operation with the vector, R applies this operation to all positions in the vector. Imagine we wanted to add 2 to all the elements in the vector, and then we would simply do:

```
> vector + 2
 [1]  3  4  5  6  7  8  9 10 11 12
```

The reader can also apply operations to vectors. For example, he/she can add another vector to the previous one. Imagine we wanted to add the following vectors:

```
> vector2 <- 3:12
> vector2
 [1]  3  4  5  6  7  8  9 10 11 12
> vector + vector2
 [1]  4  6  8 10 12 14 16 18 20 22
```

The reader must keep in mind that the vectors should have the same length. Otherwise, the compiler produces a warning and sums the vector, but it recycles the first vector to do the addition of vectors. We will return to this later, and we will explain better what happens with vectors of different lengths.

Type

The type of values that a vector can store is variable. The most used types are:

- Character,
- Logical,

- Numeric,
- Complex.

With the function *mode(),* we can check what is the type of the vector:

```
> mode(vector)
[1] "numeric"
```

As we expected, our vector object is a numeric vector with integers from 1 to 10 stored in it. An example of other types of vectors could be:

```
> char.vector <- c("String1","String2","String3")
> mode(char.vector)
[1] "character"
```

In the previous example, we used another function to create a vector, the *c()* function. This function allows us to create any vector, for instance, if we wish to create a numeric vector we could do:

```
> num.vector <- c(12.5,5.64,7.84)
> mode(num.vector)
[1] "numeric"
```

Length

Sometimes it is convenient to know the extension of the vectors. This can be achieved with the function *length()*. Some examples using this function are:

```
> length(vector)
[1] 10
> length(char.vector)
[1] 3
> length(num.vector)
[1] 3
```

Indexing

We can access the elements of a vector by using indexes. For example, to access the first element of the previously stated vector (*char.vector*) we would write the following command:

```
> char.vector[1]
[1] "String1"
```

If we would like to check a sequence of vector positions, for example 1 through 2 we can do it several ways like:

```
> char.vector[1:2]
[1] "String1" "String2"
> char.vector[c(1,2)]
[1] "String1" "String2"
```

To check, the first position of the vector char.vector and then the third, we would write it like this:

```
> char.vector[c(1,3)]
[1] "String1" "String3"
```

Vector Names

We can also name the vectors elements or positions. For example, with our vector num.vector with length 3, we could issue the following command:

```
> names(num.vector) <- c("Math Grade", "French Grade", "Ger-
man Grade")
> num.vector
  Math Grade French Grade German Grade
       12.50         5.64         7.84
```

With the previous example, it is clear now that we transformed vector num.vector with additional information about each stored element. Also, following the previous indexation procedures, we can also retrieve a vector's information through its element's name(s):

```
> num.vector["Math Grade"]
Math Grade
      12.5
> num.vector[c("Math Grade","German Grade")]
  Math Grade German Grade
       12.50         7.84
```

Logical Operations with Vectors

R allows fascinating logical operations with vectors. As an example, if we need to know the positions of the vectors with grades above ten we could do:

```
> num.vector[num.vector > 10]
Math Grade
      12.5
```

We can also do logical operations with intervals. For example, if we need to know the grades in our numeric vector that are above six AND below ten we would do:

```
> num.vector[num.vector < 10 & num.vector > 6]
German Grade
        7.84
```

Other operators like logical OR are also possible. Imagine we needed to know the grades above 10 OR below six we would do:

```
> num.vector[num.vector > 10 | num.vector < 6]
  Math Grade French Grade
       12.50         5.64
```

Functions

If the reader has been following our first examples, it is expected that he/she has already used some functions. Remember *sqrt()*, *length()*, or even *mode()*? Those are functions.

Functions are useful because they avoid the programmer to re-write all the code inside a function every time he/she wants to use it again. The great thing about new libraries or packages is that it comes generically with a set of functions that provide pre-determined operations. In other words, functions have inputs, with those inputs some internal procedures take place to give an output the user desires. Have a look at the following example of a function R code:

```
add <- function(x,y){
  x+y
}
```

This is the declaration of a function in R. The name of the function we programmed is *add()*. If we look closely, we can see that this function has two possible inputs, *x* and *y*, and the internal instruction is to add these two inputs. An example of use of this function would be:

```
> add(x=2,y=2)
[1] 4
```

Evidently, in this example, we wish to add two plus two, which are respectively the inputs *x* and *y* of the function. The result we obtain in this case is correct and equal to 4.

Statistical Functions

R has a variety of included functions that we might use in our statistical tasks. Some of them are:

- Max,
- Min,
- Mean,
- sd,
- Summary, and
- Many others.

Examples of those functions with our numeric vector num.vector would provide the following results:

```
> max(num.vector)
[1] 12.5
> min(num.vector)
[1] 5.64
> mean(num.vector)
[1] 8.66
> sd(num.vector)
[1] 3.502742
> summary(num.vector)
   Min. 1st Qu.  Median    Mean 3rd Qu.    Max.
   5.64    6.74    7.84    8.66   10.17   12.50
```

Some of these functions have names that are self-explanatory of what they do. Some others like *sd* (standard deviation) and summary will have a better explanation given further in this book (in the descriptive chapter).

Another useful function we will use later in this book is the *table()* function. Suppose we have the information about the grades of students in several Ph.D. courses. We have the following vectors:

```
> students <- c("John","Mike","Vera","Sophie","Anna","Vera"
,"Vera","Mike","Anna")
> courses <- c("Math","Math","Math","Research","Research
2","Research","Research 2","Computation","Computation")
> grades <- c(13,13,14,16,16,13,17,10,14)
```

If we want to count the number of grades we have for each student we can do:

```
> table (students)
students
  Anna    John    Mike Sophie    Vera
     2       1       2      1       3
```

Additionally, we can cross two vectors by creating a contingency table. For this we can do:

```
> table (students, courses)
courses
students Computation Math Research Research 2
  Anna             1    0        0          1
  John             0    1        0          0
  Mike             1    1        0          0
  Sophie           0    0        1          0
  Vera             0    1        1          1
```

The results show the courses each student is taking and also how many students we have for each course in this available data.

Factors

When we have character vectors, i.e. a categorical vector and a large amount of data, it is positive to store it in a compressed fashion. For example, with the vector courses, which is a character vector we can transform it to factors by using the following command:

```
> courses.factors <- factor(courses)
```

This command results in the following:

```
> courses.factors
[1] Math        Math        Math        Research    Research
2
[6] Research    Research 2  Computation Computation
Levels: Computation Math Research Research 2
```

The previous command also outputs the levels of the factor transformation. These levels are the unique values of the transformed variable.

The following function is used to check the levels of the compression of a character vector:

```
> levels(courses.factors)
[1] "Computation" "Math"        "Research"    "Research 2"
```

Data Frames

Another interesting data structure available in R is the *dataframe*. Data frames can be viewed as tables that can contain vectors of different types. For example, if we wish to transform our previous vectors students, courses, and grades to a data frame we would use the function *data.frame()* like this:

```
> my.dataframe <- data.frame(student=students,
course=courses, grade=grades)
> my.dataframe
  student       course grade
1    John         Math    13
2    Mike         Math    13
3    Vera         Math    14
4  Sophie     Research    16
5    Anna   Research 2    16
6    Vera     Research    13
7    Vera   Research 2    17
```

```
8     Mike Computation     10
9     Anna Computation     14
```

With the *data.frame()* function we can, therefore, create a dataframe with the names of each column and the respective values which in this case were our previously created vectors.

How to Edit

There is another way to create or edit a dataframe. By using the function *edit()* we can write something like the following:

```
edit(my.dataframe)
```

After inputting the previous command, a window opens in the RStudio IDE. The new window allows editing the content of the dataframe. The reader can also start a new dataframe like this. If we wanted an empty dataframe we could do:

```
> my.empty.dataframe <- data.frame()
> edit(my.empty.dataframe)
```

A new window would appear, this time, different from Figure 10. In this new window, an empty table with no values or named variables would be available for us to write values in the cells of the table. As we write the name of the variables, RStudio asks what is the type of the variable we wish to input. The options to choose are numeric or character. When we finish introducing character variables, the compiler transforms them to factors.

Indexing

There are several possible ways of reaching a value inside a dataframe structure. As an example, imagine we wanted to list all students in the dataframe. We could do it by writing down one of the following commands:

```
> my.dataframe$student
[1] John   Mike   Vera   Sophie Anna   Vera   Vera   Mike
Anna
Levels: Anna John Mike Sophie Vera
> my.dataframe[,1]
```

Figure 10. Example of an RStudio's data frame edit window

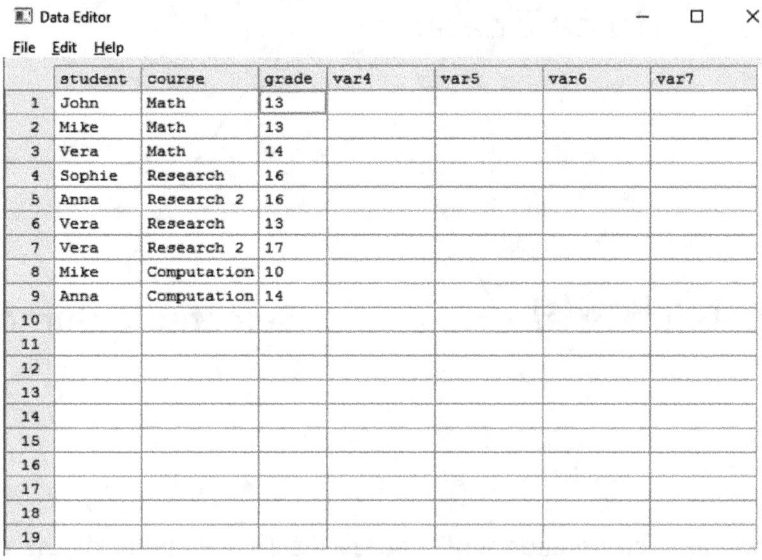

	student	course	grade	var4	var5	var6	var7
1	John	Math	13				
2	Mike	Math	13				
3	Vera	Math	14				
4	Sophie	Research	16				
5	Anna	Research 2	16				
6	Vera	Research	13				
7	Vera	Research 2	17				
8	Mike	Computation	10				
9	Anna	Computation	14				
10							
11							
12							
13							
14							
15							
16							
17							
18							
19							

```
[1] John   Mike   Vera   Sophie Anna   Vera   Vera   Mike
Anna
Levels: Anna John Mike Sophie Vera
```

In the first example, as we know the column name, we used the name of our dataframe, the symbol $ and the name of the column to check the entire column. If we did not know the name of the column, we could write down the second command, which is the basis of the indexing of dataframes. What happens inside the brackets is that the first element before the comma indicates the selected rows of the dataframe. As we can see, this is empty which means we are selecting every row. After the comma, the value 1 indicates we wish to output the column with index 1. Please verify this explanation in Figure 11.

Indexing can become even more powerful in R. As the reader might already realize, we are retrieving vectors with our last commands. If we wish to know some particular index of these vectors, we can use another index inside brackets like this:

```
> my.dataframe$student[1]
[1] John
Levels: Anna John Mike Sophie Vera
> my.dataframe[,1][1]
```

Figure 11. Schema of data frames indexing example

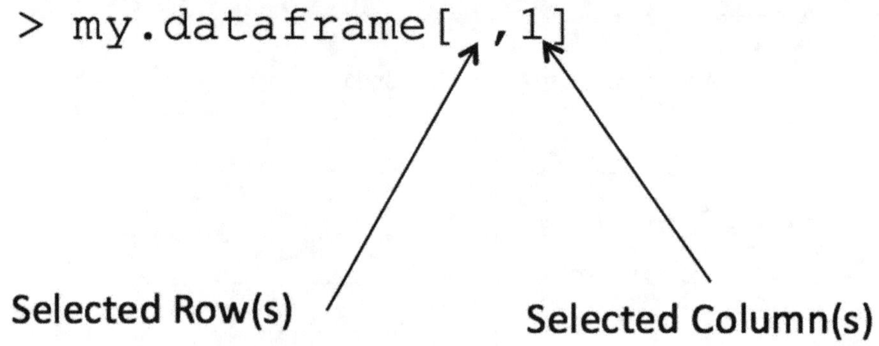


```
[1] John
Levels: Anna John Mike Sophie Vera
```

The previous commands will give us the first element of the obtained vectors.

Filters

Like we did with vectors, we can use R's powerful filtering features to extract the results we need from our dataframe. Please mind the following examples:

- Are there grades superior to 14?

```
> my.dataframe$grade > 14
[1] FALSE FALSE FALSE  TRUE  TRUE FALSE  TRUE FALSE FALSE
```

- Whose students have grades superior to 14?

```
> my.dataframe$student[my.dataframe$grade > 14]
[1] Sophie Anna   Vera
Levels: Anna John Mike Sophie Vera
```

The first command outputs either TRUE or FALSE regarding our question if there are grades superior to 14. The second command gives us the students that had these grades, superior to 14 as we wished to know.

Nonetheless, using appropriate commands can also use indexing and filtering to edit a data frame. As an example, imagine we want to change Vera's Math grade from 14 to 16. The following commands would be appropriate:

```
> my.dataframe
  student      course grade
1    John        Math    13
2    Mike        Math    13
3    Vera        Math    14
4  Sophie    Research    16
5    Anna  Research 2    16
6    Vera    Research    13
7    Vera  Research 2    17
8    Mike Computation    10
9    Anna Computation    14
> my.dataframe[3,3] <- 16
> my.dataframe
  student      course grade
1    John        Math    13
2    Mike        Math    13
3    Vera        Math    16
4  Sophie    Research    16
5    Anna  Research 2    16
6    Vera    Research    13
7    Vera  Research 2    17
8    Mike Computation    10
9    Anna Computation    14
```

Or we could use the following command if we do not know columns or row indexes:

```
> my.dataframe$grade[my.dataframe$student=="Vera" &
my.dataframe$course == "Math"] <- 16
```

If we feel a little bit lazy to write down these commands, please remember the *edit()* function we talked about before.

Useful Functions

There are some interesting functions we can use with our dataframes. Please mind the following list:

- **nrow:** Gives the dataframe's number of rows,
- **ncol:** Gives the dataframe's number of columns,
- **colnames:** Gives the dataframe's column names,
- **rownames:** Gives the dataframe's row names,
- **mode:** Gives the dataframe's data type,
- **class:** Generic function that can be used to check if it is a dataframe we are working with,
- Summary.

Examples of outputs with the previous functions are:

```
> nrow(my.dataframe)
[1] 9
> ncol(my.dataframe)
[1] 3
> colnames(my.dataframe)
[1] "student" "course"  "grade"
> rownames(my.dataframe)
[1] "1" "2" "3" "4" "5" "6" "7" "8" "9"
> mode(my.dataframe)
[1] "list"
> class(my.dataframe)
[1] "data.frame"
> summary(my.dataframe)
 student          course        grade
 Anna :2    Computation:2    Min.   :10.00
 John :1    Math       :3   1st Qu.:13.00
 Mike :2    Research   :2   Median:14.00
 Sophie:1    Research 2:2    Mean   :14.22
 Vera :3                     3rd Qu.:16.00
                              Max.   :17.00
```

Matrices

Matrices are different from dataframes in R. They can only store elements of the same type, usually numeric. They are useful to store two-dimensional data, and they can be seen as vectors of two dimensions. The function *matrix()* is appropriated to create a matrix. We use the following code to do this:

```
> my.matrix <- matrix(c(12,13,14,10,12,15,16,12), nrow=2,
ncol=4)
> my.matrix
     [,1] [,2] [,3] [,4]
[1,]   12   14   12   16
[2,]   13   10   15   12
```

The first input is the values we wish the matrix to have, the second input is the number of rows the matrix will have and the third input is the number of columns.

Nevertheless, there is an easier way to input a matrix data. For example, by using the function *data.entry()*.

With the following commands the reader will understand it better:

```
> my.matrix <- matrix(,2,4)
> data.entry(my.matrix)
```

With these commands, a new window opens. Within these window's cells, we can input the values for our 2x4 matrix. Figure 12 shows this window.

Matrix Indexing

The indexes of a matrix are identical to the data frames or vectors. They are two-dimensional. For example, keep in mind the following examples:

```
> my.matrix[1,]
[1] 12 14 12 16
> my.matrix[1,4]
[1] 16
> my.matrix[,4]
[1] 16 12
```

Figure 12. Example of an RStudio's matrix edit window

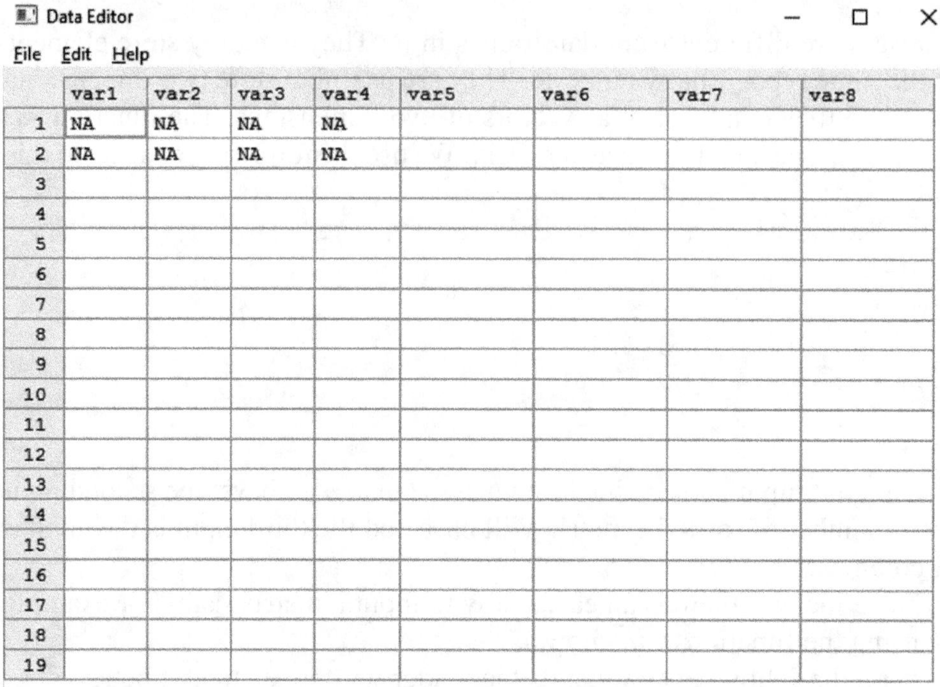

The first example would give the first row of the matrix. The second example gives the value of the first row and fourth column.

Row and Columns Names

Similar to data frames we can name columns and rows with the functions *rownames()* and *colnames()*. Please check the following examples:

```
> rownames(my.matrix) <- c("Vera","Mike")
> colnames(my.matrix) <- c("W1","W2","W3","W4")
> my.matrix
     W1 W2 W3 W4
Vera 12 14 12 16
Mike 13 10 15 12
```

Then, we can use the names we chose to retrieve values in the matrix. For example, what was Vera's grade in work 4?

```
> my.matrix["Vera","W4"]
[1] 16
```

Importing and Exporting Data with R

There are several possible ways to import data with R. We will explain one of these ways, the reading of CSV (comma separated values) files but others are also possible, like reading data from a database or an Internet URL. Later in this chapter, we will also talk how to export data to Excel.

Read CSV Files

We can read the data from a CSV file by using the function *read.csv()*. However, before opening a file with this function, we should set the working directory of R. For this, in RStudio we should look for the Session menu. Then Figure 13 clarifies where the reader should click.

After clicking Choose Directory, the user can select the directory where the CSV file is. For example, for the test.csv file with the following content:

student,course,grade
John,Math,13
Mike,Math,13
Vera,Math,14

Figure 13. Setting of the working directory of R (RStudio)

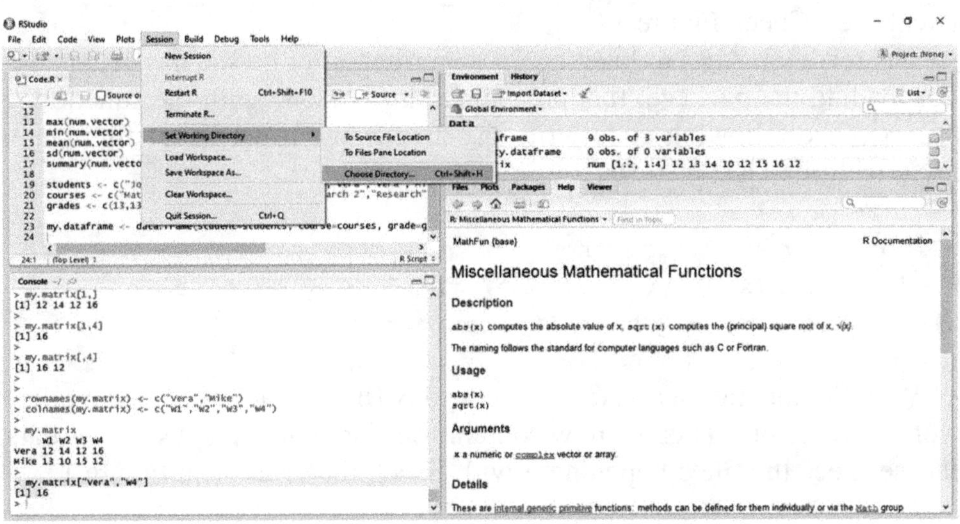

Sophie,Research,16
Anna,Research 2,16
Vera,Research,13
Vera,Research 2,17
Mike,Computation,10
Anna,Computation,14

With the following command we would read the csv file (test.csv) to a data frame named *csv.file*:

```
> csv.file <- read.csv("test.csv")
> csv.file
  student       course grade
1    John         Math    13
2    Mike         Math    13
3    Vera         Math    14
4  Sophie     Research    16
5    Anna   Research 2    16
6    Vera     Research    13
7    Vera   Research 2    17
8    Mike  Computation    10
9    Anna  Computation    14
```

Export to Excel

First, install the xlsx package. With this package, the reader can write to Excel files. Check Figure 14.

The reader just has to load the package first, after he/she has installed it. For loading the package, this procedure can be done with the function *library()*. The following code write the data frame to an Excel file named my_excel_file.xlsx:

```
> library(xlsx) #load the package
> write.xlsx(x = my.dataframe, file = "my_excel_file.
xlsx",sheetName = "Sheet 1", row.names = FALSE)
```

With the function *write.xlsx()* a new xlsx file will appear in the reader's working directory. This file now contains our familiar student's grades data. Please check the file by opening it with Excel; the result is in Figure 15.

Figure 14. Installing xlsx package in RStudio

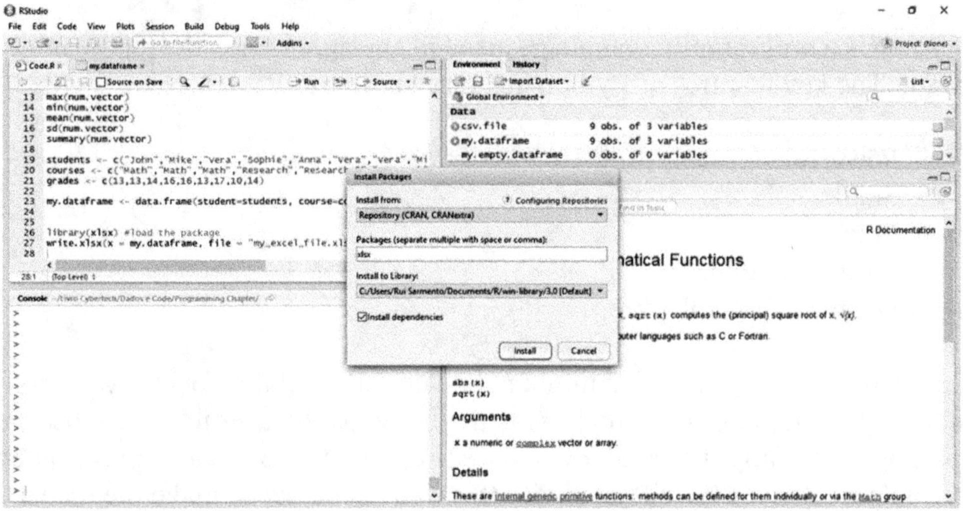

Figure 15. Example of an xlsx file opened in Excel

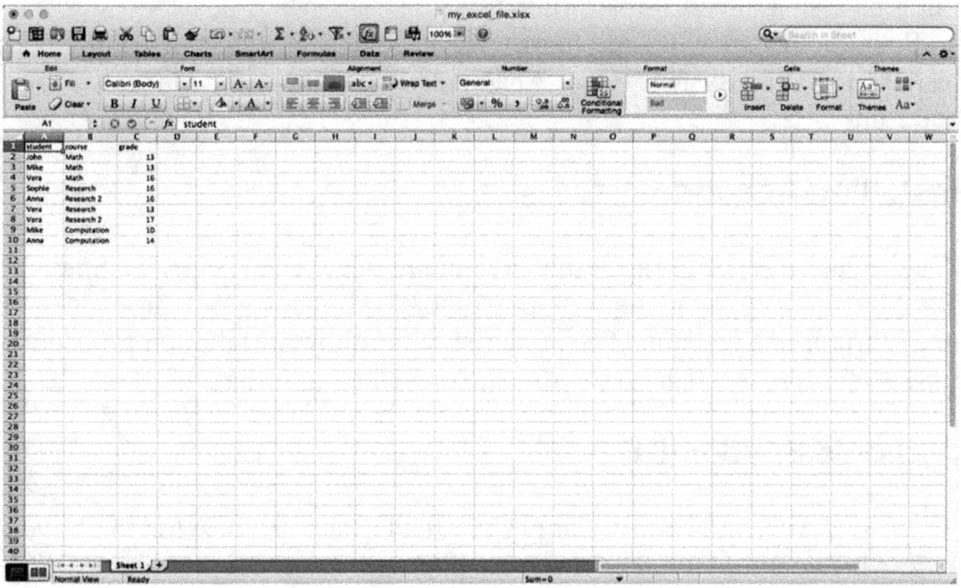

The reader might have noticed we used a new function, the *library()* function we have never used before. This function has one input, the name of the package we wish to load before using its available functions. The function we used from this package was the *write.xlsx()* function.

PYTHON

Python is a programming language, used in any application the reader might want. The key features of Python are very similar to R:

- Powerful,
- Stable,
- Free,
- Programmable,
- Open Source Software.

On the contrary side, that might not be initially suitable for everyone considering the tasks of data analysis is that it needs the user to select specific packages carefully. The reader will have to choose those that are appropriated to his/her intents. We will deal with this in this chapter to make the reader's life easier.

There are several Python distributions nowadays. Distributions are available depending on the area a language is used, and typically includes the libraries that are needed for certain tasks.

First, the reader will need to install Anacondas Python's distribution for his/her operating system (OS). Anaconda is available for Mac, Windows, and Linux.

Installing Anaconda

Anaconda is a set of libraries that are unique to the Data Analysis, Statistics and Machine Learning areas, among others. It has several libraries we will need further in this book. The reader should follow installation procedures for installing Anaconda on the website presented in Figure 16.

Python's Spyder IDE

Following Anaconda's installation, the reader should look for the Spyder IDE, which comes with the Anaconda package. This IDE provides efficient ways of working with Python and will be of great help in the tasks we have ahead in this book. Thus, we will avoid using Python's GUI and input a command at each time has we had initially to do with R GUI's and its console. The reader will immediately notice three windows on the screen as appears in Figure 17. The left window is where the commands should be written. In

Figure 16. Overview of the site to download Anaconda Python's distribution

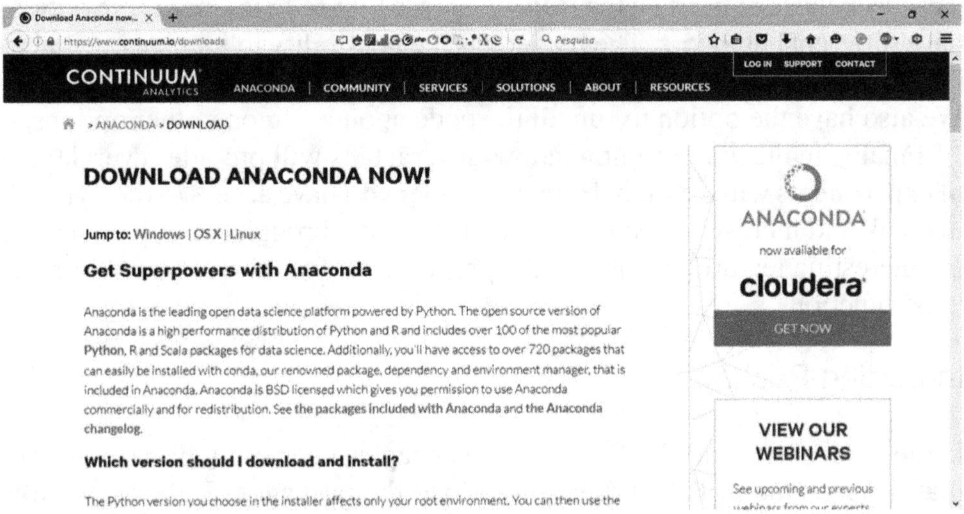

Figure 17. Spyder screen

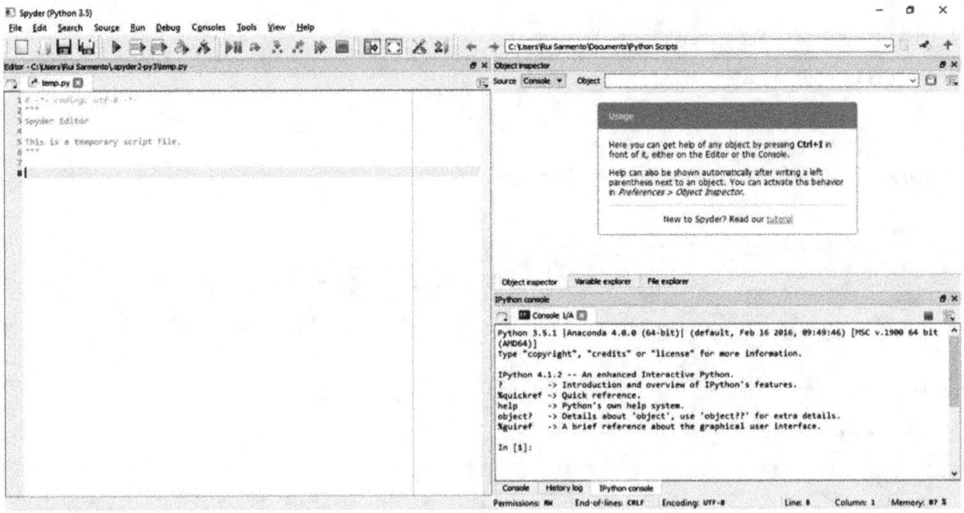

Figure 17, we inserted a similar code mentioned before when writing about R console commands.

Additionally, the reader can clearly see that, in the lower right window, the console also appears. This will be where the results of the commands appear. On the upper right window are the environment objects. The reader might be asking how to run the code by now. There are several choices; we can check those options on the Run menu (see Figure 18). Thus, these options have different behaviors. The reader has the possibility to execute one line at each time or the selected parts of the code that have selected with the mouse. We also have the option to run all the code at once, among other options.

Finally, in the upper right window, several tabs will provide several types of experiments with Python. Here, the reader will have access to the Python's manual, and inclusively can search by keywords through the guides. This is an interesting feature that allows the programmer to know more about modules' functions.

Importing Packages

If the reader has read the R part of this chapter he/she might have noticed that we used the *library()* function to load the packages. Python is similar we have to use the import keyword to load some libraries and therefore, all its available functions to use after that. For example, the reader might want to inspect the following example:

```
import math as math
x=math.pow(3,2)
```

Figure 18. Options to run the code menu

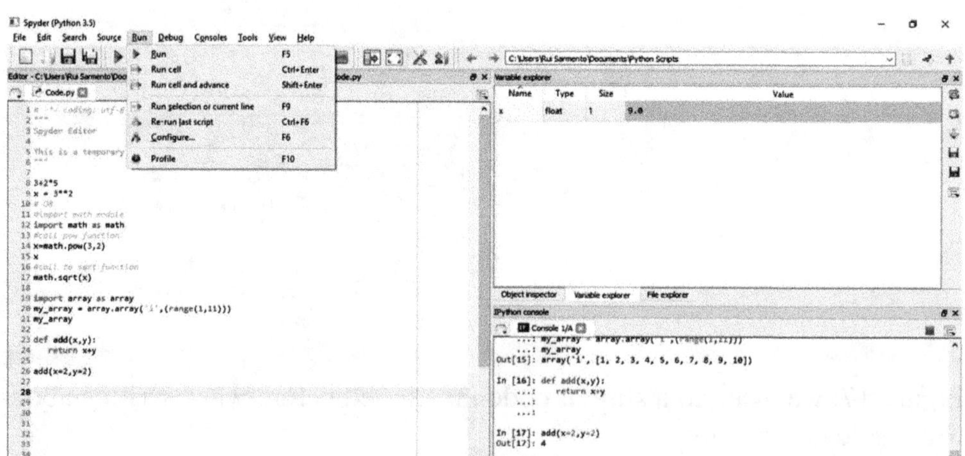

In this example, we used the keyword to *import* the math module/library. We also used the keyword *as* to name the module to a name of our choice. Thus, after this when the reader wants to call any function he/she would do it like the previous example. This time, we used the *pow()* function from math module.

Save a Variable

The reader might already acknowledge that we use the "equal" symbol to assign a value or expression to a variable. In the previous example, we assigned the expression *math.pow(3,2)* to the variable *x*.

Use a Variable in Instructions

Please check the following command:

```
#call to sqrt function
math.sqrt(x)
```

The reader might find this very similar to R language. To calculate the square root of *x*, we used the *x* variable previously set, with the previous code.

List Variables in Session

In Spyder IDE, the upper right window lists all the variables in the current session. Please check Figure 14 and remind the variable *x* is the listed variable after we have run the previous commands in this chapter.

Delete Variables

By using the powerful features of Spyder IDE, the reader can delete any variables stored in memory. Please mind Figure 15. By right clicking in the variable presented in the variable explorer, a variety of options appear. Thus, among others, the reader can select to remove the variable from memory.

Arrays

The module array defines an object type, which can compactly represent an array of basic values: characters, integers, floating point numbers. Arrays

Figure 19. Delete variables in memory (Spyder)

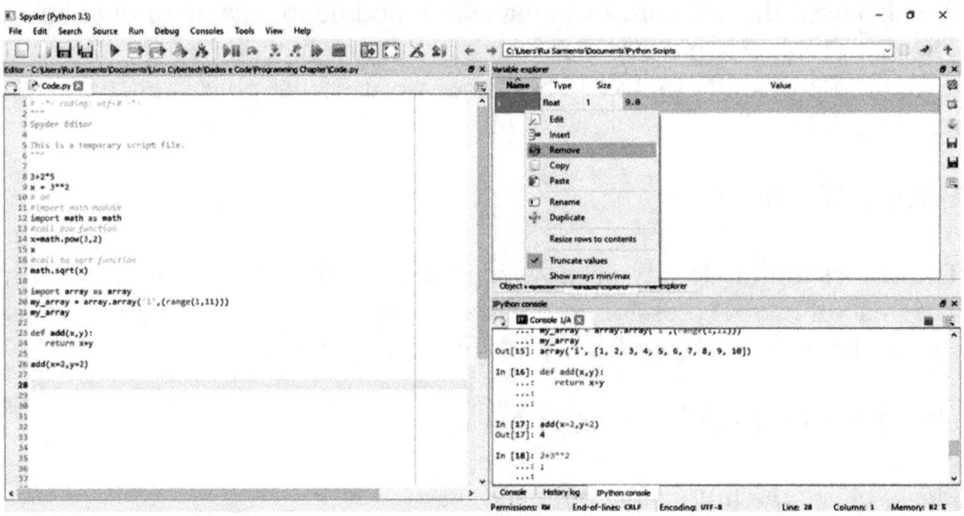

are sequence types and behave very much like lists, except that the type of objects stored in them is constrained.

To declare an array in Python, we can use the following code:

```
import array as array
my_array = array.array('i',(range(1,11)))
```

This will produce the following output:

```
...: my_array
Out[10]: array('i', [1, 2, 3, 4, 5, 6, 7, 8, 9, 10])
```

If the reader needs to do an operation with the array, Python applies a defined operation to all positions in the array. Imagine we wanted to add 2 to all elements in the array. Then, we would do:

```
[x+2 for x in my_array]
```

And the result of the operation input would be:

```
[x+2 for x in my_array]
Out[11]: [3, 4, 5, 6, 7, 8, 9, 10, 11, 12]
```

Additionally, if we wish to add two vectors we can do it in a variety of ways. Our favorite is to use the numpy module and its function *add()* like this:

```
import numpy as np
new_array = np.add(my_array, my_array2)
new_array
```

The reader might have noticed that we could apply the same function to add 2 to the array as we previously stated. With this new function we would do:

```
import numpy as np
new_array2 = np.add(my_array, 2)
new_array2
```

The result of the operation would be, as expected, similar to the previous operation with my_array.

```
Out[18]: array([ 3,  4,  5,  6,  7,  8,  9, 10, 11, 12],
dtype=int32)
```

Type

The type is specified at object creation time by using a type code input in the array function, which is a single character. There are several type possible codes. The reader should check the array module manuals for further information, as there are many possible inputs in this parameter.

Length

Sometimes it is convenient to know the extension of the arrays. This can be achieved with the function *len()*. Some examples using this function are:

```
len(my_array)
Out[18]: 10
```

Indexes

Indexation of arrays in Python is much similar with vector indexation in R. Nonetheless, keep in mind that indexes with R start in 1. In Python, the indexes of structures start with 0. Therefore, to extract the fourth position of our previously created array we would do:

```
my_array[3]
Out[19]: 4
```

Remember, due to the differences with indexation, if we were to retrieve the first position of our Python array, we would do:

```
my_array[0]
Out[20]: 1
```

Please mind the following instructions for setting an array of strings:

```
char_array = ['String1','String2','String3']
char_array
char_array[0:2]
```

We wish to output the first two elements in the array with the last command. The output would be:

```
char_array[0:2]
Out[35]: ['String1', 'String2']
```

Please keep in mind that Python has different indexation than R. The reader might have noticed that, with the previous command, we are selecting and expecting position 0 and 1 of the array. Nonetheless we declared char_array[0:2], i.e., from position 0 to position 2, excluding this last position.

If we needed to know the array value in the first position and the third we would do the following command:

```
char_array[0::2]
```

The output of this command would be:

```
char_array[0::2]
Out[38]: ['String1', 'String3']
```

Functions

The great thing about new libraries or packages is that it comes generically with a set of functions that provide pre-determined operations. In simple words, functions have inputs, and with those inputs, some internal procedures take place to give an output the user desires. Have a look at the following definition of a function Python pseudo-code:

```
def functionname(parameters):
   #intructions inside the function
   return [expression]
```

An example of a function declaration would be:

```
def add(x,y):
   return x+y
```

This is the declaration in Python of a function. Please keep in mind that Python requires respect of indentation of the code. Mind the indentation of the function declaration after the ":" signal. The name of the function we programmed is *add()*. If we look closely, we can see that this function has two possible inputs, *x* and *y*, and the private instruction is to add these two inputs. An example of the use of this function would be:

```
add(x=2,y=2)
Out[17]: 4
```

Evidently, in this example, we wish to add two plus two which are respectively the inputs *x* and *y* of the function. The result we obtain in this example is equal to 4.

Useful Functions

There are several functions we will use throughout this book that is related to data analysis and statistics. The difference to R is that the majority of those functions come included in packages directed to data and numeric analysis, statistics and others. We will explain more of those functions throughout this book and its data analysis tasks.

Dataframes

How to Create

Imagine we had the following vectors of students, courses and grades already created with Python like the following:

```
students = ["John","Mike","Vera","Sophie","Anna","Vera","Ve
ra","Mike","Anna"]
courses = ["Math","Math","Math","Research","Research
2","Research","Research 2","Computation","Computation"]
grades = [13,13,14,16,16,13,17,10,14]
```

We wish to create a data frame with these values. Therefore, we write the following commands:

```
Import pandas as pd
my_grades_dataframe = pd.concat([pd.DataFrame(students,colum
ns=['student']),pd.DataFrame(courses,columns=['course']),pd.
DataFrame(grades,columns=['grade'])], axis=1)
```

The previous command just concatenates all the arrays previously stated and after transforming each of the arrays into a data frame, by using the functions available in pandas Python's module.

How to Edit

By using Spyder's powerful IDE features, the reader can easily edit a data frame after creation. By selecting the variable explorer in the upper right window, we can right-click on the data frame we wish to edit like the Figure 20.

After clicking *edit*, the window of Figure 21 appears.

As the reader might expect, this window is very appropriate to do an edition of data frames. By selecting a cell in the table, the reader can change the values and hit the OK button. The data frame will be stored in its new version and accordingly to the reader's changes operated in the variable.

Figure 20. Editing a data frame in Spyder

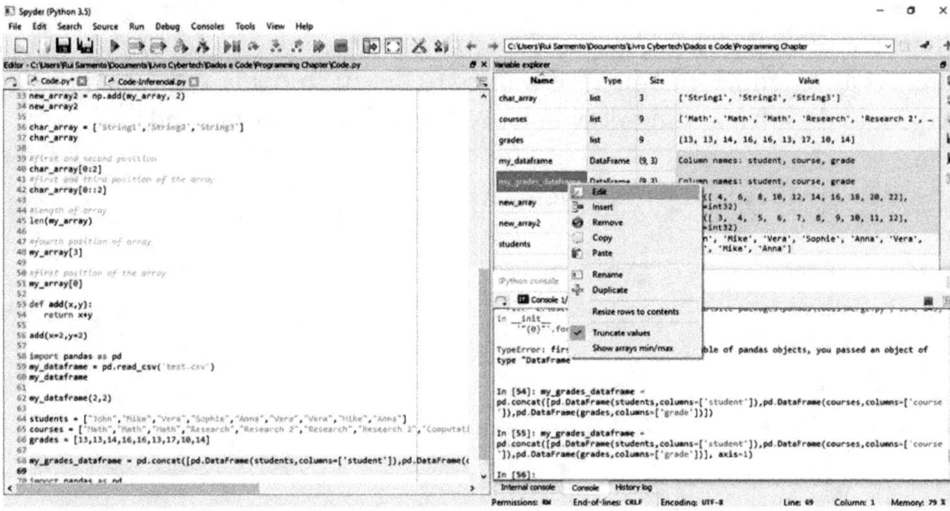

Figure 21. Edit data frame window (Spyder)

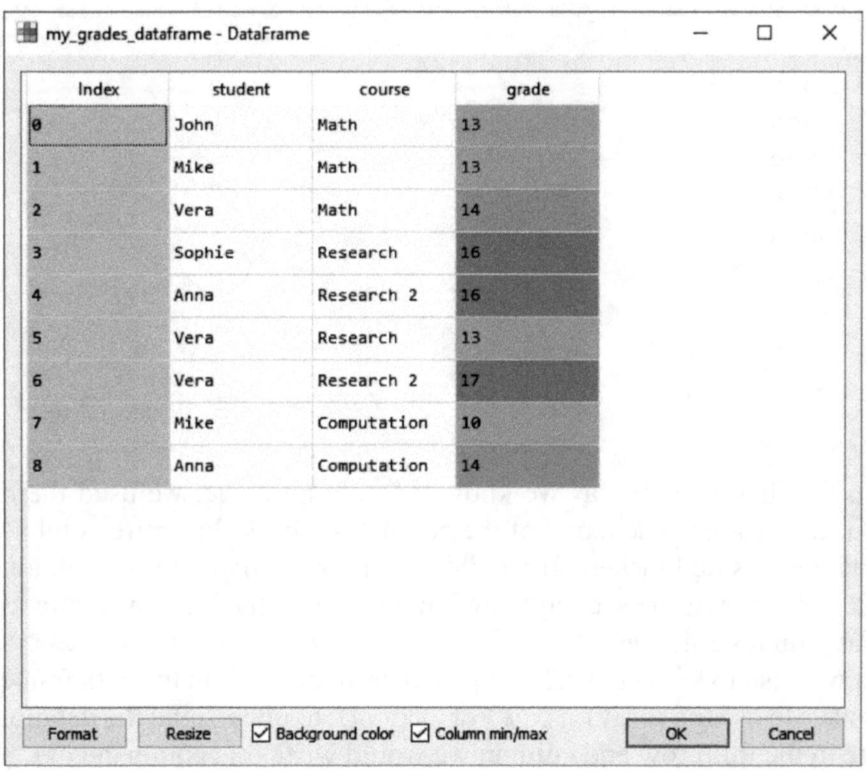

Index	student	course	grade
0	John	Math	13
1	Mike	Math	13
2	Vera	Math	14
3	Sophie	Research	16
4	Anna	Research 2	16
5	Vera	Research	13
6	Vera	Research 2	17
7	Mike	Computation	10
8	Anna	Computation	14

Indexing

There are several possible ways of reaching a value inside a data frame structure. As an example, imagine we wanted to list all students in the data frame. We could do it by writing down one of the following commands:

```
my_grades_dataframe['student']
Out[66]:
0       John
1       Mike
2       Vera
3     Sophie
4       Anna
5       Vera
6       Vera
7       Mike
8       Anna
Name: student, dtype: object
my_grades_dataframe[[0]]
Out[68]:
  student
0     John
1     Mike
2     Vera
3   Sophie
4     Anna
5     Vera
6     Vera
7     Mike
8     Anna
```

In the first example, as we know the column name, we used the name of our data frame. The name of the column to check the entire column was introduced inside brackets. If we did not know the name of the column, we could write down the second command, which is the basis of the indexing of data frames columns.

If we wish to know the cell value of a particular cell in the data frame, we will have to use the function *ix()*. For example, to understand the dataframe's value in the third row and column we would write this command:

```
my_grades_dataframe.ix[2,2]
```

The output would be Vera's Math grade, which is 14:

```
my_grades_dataframe.ix[2,2]
Out[69]: 14
```

Filters

Python's Pandas module has powerful filtering features to extract the results we need from our data frame. Please mind the following examples:

- Whose students have grades superior to 14 and to what courses?

```
#select the grades > 14
my_grades_dataframe[my_grades_dataframe['grade']>14]
Out[71]:
  student      course  grade
3  Sophie    Research     16
4    Anna  Research 2     16
6    Vera  Research 2     17
```

Nevertheless, using appropriate commands, the reader can also use indexing and filtering to edit a data frame. As an example, imagine we wish to change Vera's Math grade from 14 to 16. The following commands would be appropriate:

```
my_grades_dataframe.ix[2,2] = 16
my_grades_dataframe
Out[72]:
  student       course  grade
0    John         Math     13
1    Mike         Math     13
2    Vera         Math     16
3  Sophie     Research     16
4    Anna   Research 2     16
5    Vera     Research     13
6    Vera   Research 2     17
7    Mike  Computation     10
8    Anna  Computation     14
```

If we feel a little bit lazy to write down these commands, please remember that the reader can edit the data frame with the Spyder's editing feature we talked about before.

Useful Functions

There are some useful functions regarding data frames with Python. For example, the *info()* function retrieves, among other information, the number of rows, columns and the memory usage of the data structure:

```
my_grades_dataframe.info()
<class 'pandas.core.frame.DataFrame'>
RangeIndex: 9 entries, 0 to 8
Data columns (total 3 columns):
student     9 non-null object
course      9 non-null object
grade       9 non-null int64
dtypes: int64(1), object(2)
memory usage: 296.0+ bytes
```

Pandas DataFrame's also have a *describe* method, which is ideal for seeing basic statistics about the dataset's numeric columns. For example, with the following code:

```
my_grades_dataframe.describe()
Out[76]:
           grade
count   9.000000
mean   14.222222
std     2.223611
min    10.000000
25%    13.000000
50%    14.000000
75%    16.000000
max    17.000000
```

Matrices

Matrices with Python are also possible. The reader should use *numpy* package to be able to create a matrix with a simple procedure. Please mind the following example:

```
my_matrix = np.matrix('0 0 0 0; 0 0 0 0')
my_matrix
Out[78]:
matrix([[0, 0, 0, 0],
        [0, 0, 0, 0]])
```

This time, we created a 2x4 matrix of zeros.

Insert Data in a Matrix

Using Spyder's editing options as previously stated we could change the values in the matrix. Please check the Figure 22.

Imagine we wish to change the value of the second line and fourth column; then we would have Figure 23.

We have changed the value to 5 as the previous figure represents. Nonetheless, we could also do it by writing the following command:

Figure 22. Editing matrices in Spyder

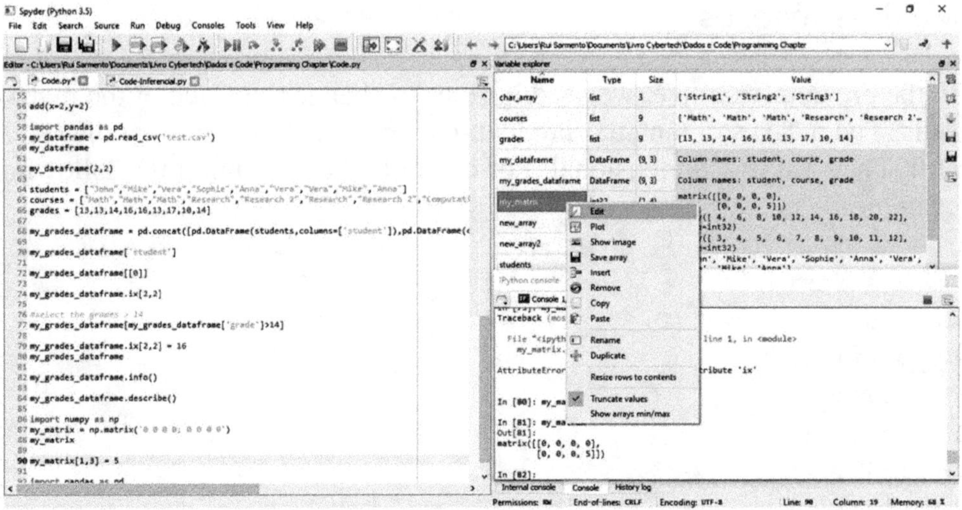

Figure 23. Matrix edition window (Spyder)

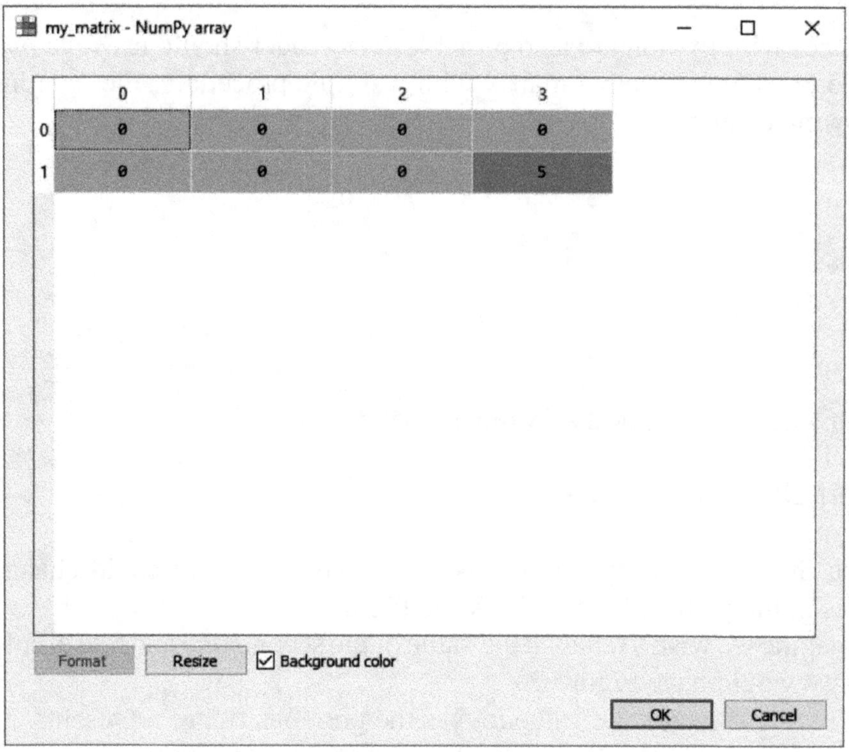


```
my_matrix[1,3] = 5
```

Matrices Indexes

In the previous example, we used indexes to change the value of the matrix cells. The indexes of a matrix are identical to the dataframes, and they start at 0. They are two-dimensional. For example, keep in mind the following examples:

```
 my_matrix[0,]
Out[83]: matrix([[0, 0, 0, 0]])
my_matrix[0,3]
Out[84]: 0
my_matrix[:,3]
Out[94]:
matrix([[0],
        [5]])
```

The first example would give the first row of the matrix. The second example gives the value of the first row and fourth column. The third example will give the reader all the values of the fourth column.

Importing and Exporting Data with Python

Read CSV Files

Reading data from CSV files is also a great feature of Python. We can obtain a data frame from a CSV.

For example, for the test.csv file consider the following content:

```
student,course,grade
John,Math,13
Mike,Math,13
Vera,Math,14
Sophie,Research,16
Anna,Research 2,16
Vera,Research,13
Vera,Research 2,17
Mike,Computation,10
Anna,Computation,14
```

First, before reading the previous data from a file, it is necessary to change the working directory to the directory where our test.csv file is. To do this, please check Figure 24. We can browse a working directory in the folder icon in the upper right corner of the Spyder IDE.

Then, with the following code, it is possible to import the data to the data frame:

```
import pandas as pd
my_dataframe = pd.read_csv('test.csv')
my_dataframe
Out[26]:
  student       course  grade
0    John         Math     13
1    Mike         Math     13
2    Vera         Math     14
3  Sophie     Research     16
4    Anna   Research 2     16
```

Figure 24. Changing the working directory (Spyder)

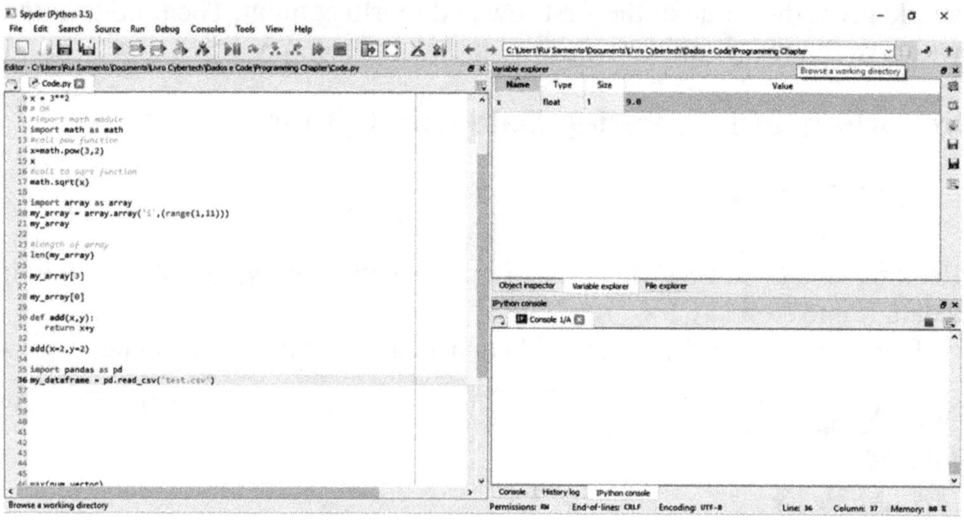


```
5     Vera       Research      13
6     Vera     Research 2      17
7     Mike   Computation       10
8     Anna   Computation       14
```

Export to Excel

The Python's package named pandas has a great function for this task. The function *to_excel* provides a way to store data frames in Excel files. The following command:

```
import pandas as pd
my_dataframe.to_excel('my_excel_file_python.xlsx', sheet_
name='Sheet1')
```

will provide an excel file named 'my_excel_file_python.xlsx'. The result is represented in Figure 25.

Connecting to Other Languages

The Python's versatility as a generic language allows the use of other languages within its programming instructions. One of these possible languages is R. Further in this book we will use this Python's feature to execute and exemplify

Figure 25. Excel file output (Python)

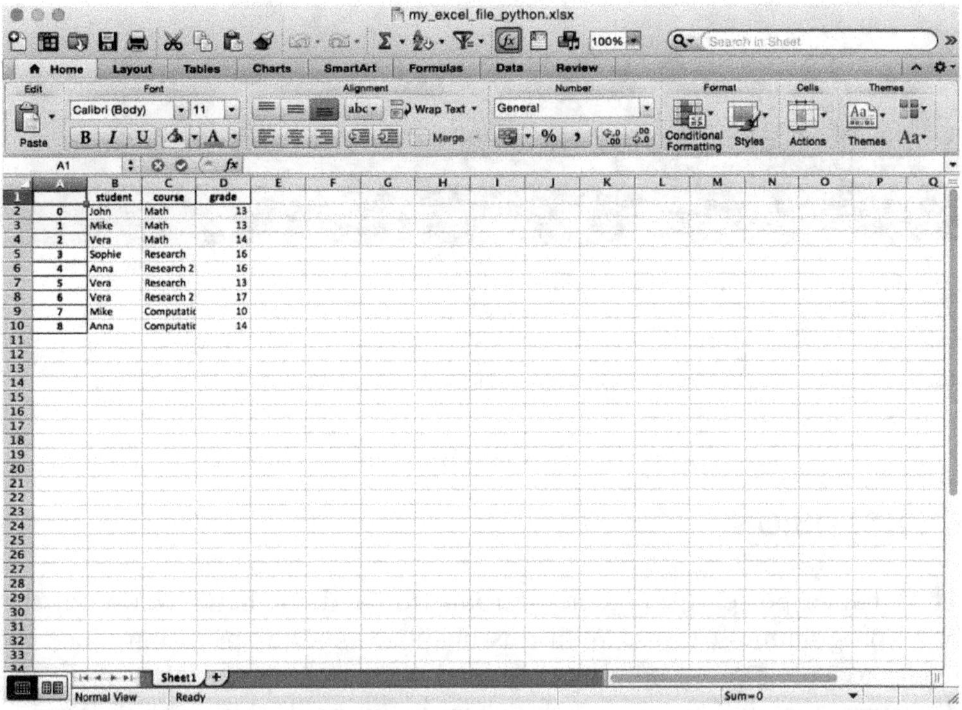

some statistical tasks. The *rpy2* module delivers just what is expected from a connection with another language, specifically R language.

To proceed with the installation of this package, some installation stages are necessary, and the reader should also install R on his/her computer. Then, the reader should download the package for his/her OS. With windows, the packages are available on a website. The selected.*whl (rpy2-2.8.1-cp35-cp35m-win_amd64.whl)* file was appropriated for the installed Python version and 64bit Windows version.

Then, within the Anaconda's console the following command was inputted:

```
pip install rpy2-2.8.1-cp35-cp35m-win_amd64.whl
```

Figure 26 illustrates the input of the previous command and the successful installation of the package *rpy2* in its version 2.8.1.

Following the installation procedure, the usual importation of the new module is now possible. For example, to call the new module in a piece of code, the programmer would write:

Figure 26. Installing rpy2 module package in Python

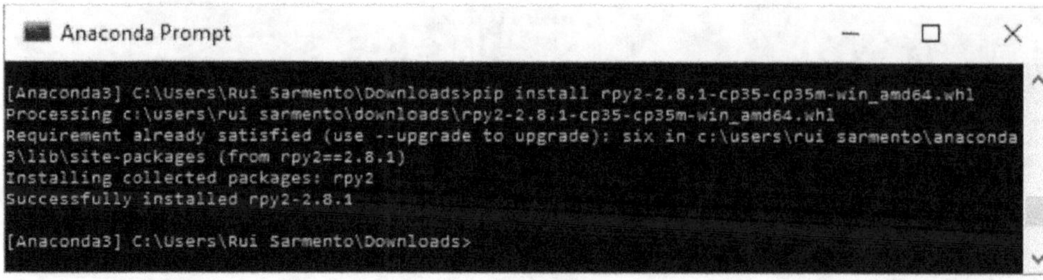


```
import rpy2 as rpy2
from rpy2.robjects.packages import importr
import rpy2.robjects as ro
```

CONCLUSION

This chapter presents a reader's introduction and contextualization of this book programming tasks. With this chapter, we attempt to introduce the reader to simple programming tasks and his/her comprehension of the use of features that will be applied elsewhere in this book. Although the book is organized with a crescent complexity of materials, the reader will encounter an imminently practical book with examples throughout.

Additionally, in this chapter, we provided a brief summary of the syntax of the languages we are focusing. We introduced the reader to their consoles, GUIs and IDEs, either for R or Python. We stress that, in this chapter, we approached just a little bit of the existing material regarding both programming languages. Nonetheless, we believe that it is possible for the reader to gather information from other sources and we tried to state them also in this chapter. Consultation of manuals and another information as we go is a required and needed procedure when learning programming languages.

We will further explore both languages, and the reader will gain a broader look into programming and statistics at the end of the following chapters.

The key concepts presented in this chapter include the programming of:

- Vectors,
- Dataframes,
- Matrices,
- Functions.

And the installation of:

- R,
- RStudio,
- Anaconda Python's Distribution.

Additionally, the reader learned basic operation concepts with both languages IDE's, RStudio for R and Spyder for Python.

Chapter 3
Dataset

INTRODUCTION

In this chapter, we present the dataset used in the course of this book. The dataset is composed of several variables of different types. The variables also have different distributions.

Our case study is built upon fictional data "collected" from a group of 200 data analysts. The "survey" implied collecting data like the age, gender, Python and R languages usage and the number of scientific publications per individual. Additionally, we registered what was the primary task of each researcher. We will now explain each of the variables with more detail.

VARIABLES

All variables were generated following specific constraints that could provide a broader look at statistical analysis through their characteristics variability. Therefore, this approach enables a large type of possible example analysis the reader can find throughout the book.

- **id:** "id" is a numerical type variable that provides identification of an individual. Its value is unique for each individual covered under the universe of the dataset.
- **Age:** "Age" is a numerical type of variable providing the current age of each individual.

DOI: 10.4018/978-1-68318-016-6.ch003

- **Gender:** "Gender" is a nominal variable providing the gender of each individual. This variable has two possible values, "Male" or "Female".
- **Python_user:** "Python_user" is a nominal type of variable and has the information if the individual is a frequent user of Python language in his or her data analysis task. This variable has two possible values, "Yes" or "No".
- **R_user:** "R_user" is a nominal type of variable and has the information if the individual is a frequent user of R software in his or her data analysis task. This variable has two possible values, "Yes" or "No".
- **Publications:** "Publications" is a numeric type of variable and its value indicates the number of publications the individual data analyst made until the date of the data collection.
- **Tasks:** "Tasks" is a nominal type of variable and its values indicate the position or functions the researcher performs in his institution. The three possible values for this variable are:
  - Phd Student,
  - Postdoctoral Research,
  - PhD Supervisor.
- **Q1 to Q10:** Variables "Q1" to "Q10" are the results of a survey's questionnaire presented to our researchers, subjected to this study. The presented survey was:
  - **Q1:** I feel that research tools (software, hardware, books, and others) I currently use are enough to achieve my research goals.
  - **Q2:** I understand that my research area provides the opportunity to achieve excellent productivity (published papers, book chapters, books, etc.).
  - **Q3:** My scientific productivity increased in the last year.
  - **Q4:** I feel I can improve some of my research methods.
  - **Q5:** My research methods changed very much with time.
  - **Q6:** I quickly adapted to new research tools throughout time when I needed.
  - **Q7:** I am receptive to learn new research tools that might appear in the future.
  - **Q8:** I am sure my research methods are directly related to my scientific productivity.
  - **Q9:** I would change my research tools if I were given a chance to do that.
  - **Q10:** I feel that my research tools improved in the last few years.

The researchers were then asked to classify each of the questionnaire statement with a Likert scale, which is defined as:

- 1 – "Strongly Disagree",
- 2 – "Disagree",
- 3 – "Neutral",
- 4 – "Agree",
- 5 – "Strongly Agree".

- **Year:** "Year" is a numeric type of variable and its values indicate the year the researcher published his greater amount of publications, i.e. the year with highest publishing productivity.

PRE-PROCESSING

Dealing with data is a task that frequently requires some procedures to prepare it for analysis. Pre-processing of data is a necessary task to adapt the data to the needs of the analyst. For example, either changing types of raw data variables after reading it from a CSV file, or modifying the name of the variables of the obtained dataset, among others. Many tasks are possible in pre-processing, and we will deal with a few in this chapter as they were used throughout this book.

Pre-Processing in R

R language has several functions adequate to do pre-processing of data. For example, if the programmer needs to change the codification of categorical variables to binary or even remove NA values from the data, he/she could use the following code and functions in Table 1.

Pre-Processing in Python

Similarly to R, Python language has several modules and functions adequate to do pre-processing of data. For example, if the programmer needs to change the codification of categorical variables to binary or even remove NA values from the data, he/she could use the following code and functions in Table 2.

From the previous outputs, it is evident we coded the variable R_user with a different type. Now, the variable is binary. This is useful for example to retrieve regression models as the reader might find in a subsequent chapter.

Table 1. R language: example code and functions

	In R

Code	`#remove line with NA’s`
	`data<-na.omit(data`
	`#Replace values`
	`data$Gender<-ifelse(data$Gender==“female”, 1, 0)`
	`#Replace values`
	`data$Python_user<-ifelse(data$Python_user==“yes”, 1, 0)`
	`#Replace values`
	`data$R_user<-ifelse(data$R_user==“yes”, 1, 0)`
	`i) #Output example sample before pre-processing`
	`data[1:10, “R_user”]`
	`ii) #Output example sample after pre-processing`
	`data[1:10, “R_user”]`

Output	[1] yes yes no no no no no no no no Levels: no yes [1] 1 1 0 0

CONCLUSION

This chapter is small, but it is not less important. Here we described the dataset used throughout the book. The dataset has different types of variables that will be employed, further in this book, to exemplify some of the statistical tasks the reader might want to perform with his dataset. It also sets the context of this book and explains that we choose the academic research theme. The theme is just arbitrary and all content of this book; its approached statistical tasks are also applicable to any dataset the reader might want to explore.

Succinctly, in this chapter we addressed:

- Dataset Variables,
- Variable’s Types,
- Pre-processing in R (Introduction),
- Pre-processing in Python (Introduction).

Table 2. Python language: example code and functions

<table>
<tr><th></th><th>In Python</th></tr>
<tr><td>Code</td><td><pre>import numpy as np
#remove line in data where Age=NaN
datadf = datadf[np.isfinite(datadf['Age'])]
#remove line in data where Python_user=NaN
datadf = datadf.dropna(subset=['Python_user'])
#Replace values
datadf['Gender'] = datadf['Gender'].replace(['male','female'],[0,1])
#Replace values
datadf['Python_user'] = datadf['Python_user'].
replace(['no','yes'],[0,1])
#Replace values
datadf['R_user'] = datadf['R_user'].replace(['no','yes'],[0,1])
i) #Output example sample before pre-processing
datadf.ix[0:9,['R_user']]
ii) #Output example sample after pre-processing
datadf.ix[0:9,['R_user']]</pre></td></tr>
<tr><td>Output</td><td><pre>Out[5]:
R_user
0 yes
1 yes
2 no
3 no
4 no
5 no
6 no
7 no
8 no
9 no
Out[7]:
R_user
0 1
1 1
2 0
3 0
4 0
5 0
6 0
7 0
8 0
9 0</pre></td></tr>
</table>

Chapter 4
Descriptive Analysis

INTRODUCTION

Descriptive statistics is the initial stage of analysis used to describe and summarize data. The availability of a large amount of data and very efficient computational methods strengthened this area of the statistic.

R VS. PYTHON

In order to make a descriptive analysis correctly, the first step should be to identify the variable type. In the following sections, suggestions for this analysis are presented, in R and Python languages.

Categorical Variables

Categorical variables are qualitative variables and cannot be represented by a number. In the data set in the study, the Gender of the researchers, R_user (yes vs. no) and Python_user (yes vs. no) are categorical variables. In particular, they are nominal categorical variables. The difference between nominal and ordinal categorical variables is that, regarding their presentation, while nominal variables may be presented randomly or in the preferred order of the analyst, the ordinal variables must be presented in the order that is more easily understood. For example, the school levels must be shown by order, i.e., from the lowest level to the highest.

DOI: 10.4018/978-1-68318-016-6.ch004

In case of categorical variables, the analysis that can be done is the frequency of each category. In R, this count is given by the *table* function. In Python, the *value_counts()* function gives these values. A suggestion of the descriptive analysis for the Gender variable, in the programming languages mentioned above are shown in Table 1 and Table 2.

Table 1. R language: a suggestion of the descriptive analysis of Gender

	In R

Code	`### Gender’s descriptive analysis`
	`# Count of each factor level of the Gender variable (presented in a data frame “data.df” with a “Gender” #column), and conversion of the count to numeric values`
	`Freq.Gender <- as.numeric(table(data.df[,”Gender”]))`
	`# Cumulative frequencies of the “Freq.Gender” object`
	`CFreq.Freq.Gender <- cumsum(Freq.Gender)`
	`# Relative frequencies for each factor level of the Gender variable and conversion of the frequencies as numeric values`
	`Rel.Freq.Gender <- as.numeric(prop.table(Freq.Gender))`
	`# Data frame with the realized analysis`
	`Freqs.Gender <- data.frame(Gender = levels(factor(data.df[,”Gender”])), Frequency = Freq.Gender, Cumulative.Frequency = CFreq.Freq.Gender, Relative.Frequency = Rel.Freq.Gender)`
	`# Output the previous results`
	`Freqs.Gender`

Output	`  Gender  Frequency  Cumulative.Frequency  Relative.Frequency`
	`1 female  87         87                    0.435`
	`2 male    113        200                   0.565`

Table 2. Python language: a suggestion of the descriptive analysis of Gender

	In Python

Code	`### Gender descriptive analysis`
	`# Count of each factor level of the Gender variable`
	`print(datadf[‘Gender’].value_counts())`
	`# Filtering Gender data`
	`gender_datadf = datadf[‘Gender’]`
	`# Group by Gender value`
	`gender_datadf = pd.DataFrame(gender_datadf.value_counts(sort=True))`
	`# Create new column with cumulative sum`
	`gender_datadf[‘cum_sum’] = gender_datadf[‘Gender’].cumsum()`
	`# Create new column with relative frequency`
	`gender_datadf[‘cum_perc’] = 100*gender_datadf[‘cum_sum’]/gender_datadf[‘Gender’].sum()`
	`gender_datadf`

Output	`Out[130]:`
	`Gender  cum_sum  cum_perc`
	`male    113 113  56.5`
	`female  87 200   100.0`

From the previous outputs (frequency column in R or Gender column in Python), there are 87 female and 113 male cases in the study, corresponding to 43.5% and 56.5% (relative frequencies column in R or cum_perc column in Python) respectively. The cumulative frequencies column shows that, in total, there are 200 elements in the study. The mode (most common element) of this variable is "male".

To analyze the Python_user variable (a categorical variable), a similar study to the study of the Gender variable was also done. See Table 3 and Table 4.

Table 3. R language: a similar study to the study of the Gender variable

	In R

Code	`### Python_user descriptive analysis`
	`# Count of each factor level of the Python_user variable and conversion of the count as numeric values`
	`Freq.Python <- as.numeric(table(data.df[,"Python_user"]))`
	`# Cumulative frequencies of the "Freq.Python" object`
	`CFreq.Freq.Python <- cumsum(Freq.Python)`
	`# Relative frequencies for each factor level of the Python_user variable and conversion of the frequencies as numeric values`
	`Rel.Freq.Python <- as.numeric(prop.table(Freq.Python))`
	`# Data frame with the executed analysis`
	`Freqs.Python <- data.frame(Python.user = levels(factor(data.df[,"Python_user"])), Frequency = Freq.Python, Cumulative.Frequency = CFreq.Freq.Python, Relative.Frequency = Rel.Freq.Python)`
	`Freqs.Python`

Output	`  Python.user Frequency Cumulative.Frequency Relative.Frequency`
	`1 no          92        92                   0.4623116`
	`2 yes         107       199                  0.5376884`

Table 4. Python language: a similar study to the study of the Gender variable

	In Python

Code	`### Python_user descriptive analysis`
	`# Filtering Python_user data`
	`python_datadf = datadf['Python_user']`
	`# Group by Python_user`
	`python_datadf = pd.DataFrame(python_datadf.value_counts(sort=True))`
	`# Create new column with cumulative sum`
	`python_datadf['cum_sum'] = python_datadf['Python_user'].cumsum()`
	`# Create new column with relative frequency`
	`python_datadf['cum_perc'] = 100*python_datadf['cum_sum']/python_datadf['Python_user'].sum()`
	`python_datadf`
	`Freqs.Python`

Output	`Out[131]:`
	`    Python_user  cum_sum  cum_perc`
	`yes         107      107   53.768844`
	`no           92      199  100.000000`

As it is possible to observe in the previous output, the Python programming language has 107 users. The number of non-users is 92. So, as the total number of researchers (Python users and non-users) is 199, the conclusion is that there is one missing.

To calculate the relative frequency, in R, the *prop.table* function is used. The function ignores missing values (or NA's). Thus, there are 46.2% of non-users and 53.8% of users. To include NA as a category in counts, the option *exclude=NULL* should be included in the *table* function. In Python, by default, the NA's are also excluded. To include NA in the counts, in function *value_counts*, the option *dropna =False* should be added. Thus, the previous code should be replaced by the next. See Table 5 and Table 6.

Table 5. Using the prop.table function

	In R

Code	`### Python_user descriptive analysis`
	`Freq.Python <- as.numeric(table(data.df[,"Python_user"],`
	`exclude=NULL))`
	`CFreq.Freq.Python <- cumsum(Freq.Python)`
	`Rel.Freq.Python <- as.numeric(prop.table(Freq.Python))`
	`Freqs.Python <- data.frame(Python.user = levels(factor(data.`
	`df[,"Python_user"], exclude = NULL)), Frequency = Freq.Python,`
	`Cumulative.Frequency = CFreq.Freq.Python, Relative.Frequency = Rel.`
	`Freq.Python)`
	`Freqs.Python`

Output	`  Python.user Frequency Cumulative.Frequency Relative.Frequency`
	`1 no          92        92                   0.460`
	`2 yes         107       199                  0.535`
	`3 <NA>        1         200                  0.005`

Table 6. Using the value_counts function

	In Python

Code	`### Python_user descriptive analysis`
	`print(datadf['Python_user'].value_counts())`
	`python_datadf = datadf['Python_user']`
	`python_datadf = pd.DataFrame(python_datadf.value_counts(sort=True,`
	`dropna =False))`
	`python_datadf['cum_sum'] = python_datadf['Python_user'].cumsum()`
	`python_datadf['cum_perc'] = 100*python_datadf['cum_sum']/python_`
	`datadf['Python_user'].sum()`
	`python_datadf`

Output	`Out[8]:`
	`   Python_user  cum_sum  cum_perc`
	`yes        107      107      53.5`
	`no          92      199      99.5`
	`NaN          1      200     100.0`

With the previous code, the number of missing values and corresponding relative frequency is given. Thus, the Python_user variable has one missing, corresponding to 0.5% of the sample. Also, there are 92 non-users (46%) and 107 users (53.5%) of Python.

The mode (most frequent element) of this variable is "yes".

Similar to the previous variables, the analysis of the R_user variable could be done as presented in Table 7 and Table 8.

Table 7. R language: analysis of the R_user variable

	In R

Code	`### R_user descriptive analysis`
	`# Count of each factor level of the R_user variable`
	`Freq.R <- as.numeric(table(data.df[,"R_user"]))`
	`# Cumulative frequencies of the "Freq.R" object`
	`CFreq.Freq.R <- cumsum(Freq.R)`
	`# Relative frequencies of each factor level of the R_user variable and conversion of the frequencies as numeric values`
	`Rel.Freq.R <- as.numeric(prop.table(Freq.R))`
	`# Data frame with the realized analysis`
	`Freqs.R <- data.frame(R.user = levels(factor(data.df[,"R_user"])), Frequency = Freq.R, Cumulative.Frequency = CFreq.Freq.R, Relative.Frequency = Rel.Freq.R)`
	`Freqs.R`

Output	`  R.user Frequency Cumulative.Frequency Relative.Frequency`
	`1     no        91                   91              0.455`
	`2    yes       109                  200              0.545`

Table 8. Python language: analysis of the R_user variable

	In Python

Code	`### R_user descriptive analysis`
	`print(datadf['R_user'].value_counts())`
	`# Filtering R_user data`
	`r_datadf = datadf['R_user']`
	`# Group by R_user`
	`r_datadf = pd.DataFrame(r_datadf.value_counts(sort=True))`
	`# Create new column with cumulative sum`
	`r_datadf['cum_sum'] = r_datadf['R_user'].cumsum()`
	`# Create new column with relative frequency`
	`r_datadf['cum_perc'] = 100*r_datadf['cum_sum']/r_datadf['R_user'].sum()`
	`r_datadf`

Output	`Out[132]:`
	`     R_user  cum_sum  cum_perc`
	`yes     109      109      54.5`
	`no       91      200     100.0`

Regarding the R_user variable, there are 109 users (54.5%) and 91 non-users (45.5%). The mode (most frequent element) of this variable is "yes".

Regarding the individual's tasks as seen above in Table 9 and Table 10, there are 78 Ph.D. Students, 56 Ph.D. Supervisors, and 66 Postdoc researchers, corresponding to 39%, 28%, and 33%, respectively.

Table 9. R language: "Tasks"

	In R

Code	`### Tasks descriptive analysis`
	`# Count of each factor level of the variable Tasks`
	`Freq.Tasks <- as.numeric(table(data.df[,"Tasks"]))`
	`# Cumulative frequencies of the "Freq.Tasks" object`
	`CFreq.Freq.Tasks <- cumsum(Freq.Tasks)`
	`# Relative frequencies each factor level of the R_user variable and conversion of the frequencies as numeric values`
	`Rel.Freq.Tasks <- as.numeric(prop.table(Freq.Tasks))`
	`# Data frame with the realized analysis`
	`Freqs.Tasks <- data.frame(Tasks = levels(factor(data.df[,"Tasks"])), Frequency = Freq.Tasks, Cumulative.Frequency = CFreq.Freq.Tasks, Relative.Frequency = Rel.Freq.Tasks)`
	`Freqs.R`

Output	`  Tasks                  Frequency  Cumulative.Frequency  Relative.Frequency`
	`1 PhD_Student               78         78                     0.39`
	`2 Phd_Supervisor            56         134                    0.28`
	`3 Postdoctoral_research 66            200                    0.33`

Table 10. Python language: "Tasks"

	In Python

Code	`### Tasks descriptive analysis`
	`# Filtering Tasks data`
	`tasks_datadf = datadf['Tasks']`
	`# Group by tasks`
	`tasks_datadf = pd.DataFrame(tasks_datadf.value_counts(sort=True))`
	`# Create new column with cumulative sum`
	`tasks_datadf['cum_sum'] = tasks_datadf['Tasks'].cumsum()`
	`# Create new column with relative frequency`
	`tasks_datadf['cum_perc'] = 100*tasks_datadf['cum_sum']/tasks_datadf['Tasks'].sum()`
	`tasks_datadf`

Output	`Out[133]:`
	`Tasks                    cum_sum  cum_perc`
	`PhD_Student              78 78    39.0`
	`Postdoctoral_research 66 144   72.0`
	`Phd_Supervisor           56 200   100.0`

Discrete Numerical Variables

As mentioned in the Statistics chapter, a discrete numeric variable only has distinct integer values. In this case, information such as mean, standard deviation, quartiles and median, are crucial since they show the general behavior of the variable in the study. In this context, this type of descriptive analysis was applied to Age and Publication variables.

In R, some commands can be used, namely, *mean*, *median*, *min*, *max*, *quantile (x, 0.25)*, *quantile (x, 0.75)* and *sd*. However, the *summary* function gives all these values. See Table 11.

Thus, if the reader only needs the mean (for example), he/she can use the corresponding command. But, if he/she need all information about the variable, the *summary* function is more convenient. Please note that the function does not give the standard deviation of the numeric variable.

In Python, each measure should be calculated individually. Some functions like *len()*, *min()*, *max()*, *mean()*, *var()*, and *std()* are very useful. See Table 12.

Table 11. R language: age and publication variables

	In R

Code	`### Age and number of publications descriptive analysis`
	`# Summary description of Age and Publications variables, selecting the columns by the corresponding name`
	`summary(data.df[,c("Age","Publications")])`
	`OR`
	`# Summary description of Age and Publications variables, selecting the columns by the corresponding position, i.e., fifth and sixth column`
	`summary(data.df[,c(5,6)])`
	`# Standard deviation of Age variable, removing missing values, represented by NA's`
	`sd(data.df[,"Age"], na.rm=TRUE)`
	`# Standard deviation of Publications variable, removing missing values, represented by NA's`
	`sd(data.df[,"Publications"], na.rm=TRUE)`

Outputs	`# Summary of Age and Publications variables`
	`Age Publications`
	`Min. :24.00 Min. :11.00`
	`1st Qu.:33.00 1st Qu.:25.00`
	`Median:37.00 Median:31.00`
	`Mean :37.06 Mean :29.65`
	`3rd Qu.:41.00 3rd Qu.:35.00`
	`Max. :52.00 Max. :70.00`
	`NA's :2`
	`# Standard deviation of Age variable and Publications variable`
	`[1] 5.637253`
	`[1] 7.743209`

Table 12. Python language: age and publication variables

	In Python

Code	`### Age and number of publications descriptive analysis`
	`# Import panda package`
	`import pandas as pd`
	`# Read data`
	`datadf = pd.read_csv('data.csv', sep=',')`
	`## Age`
	`# To write the name of the output list`
	`print("\nAge Variable: \n")`
	`# Dimension of the Age variable`
	`print("Number of elements: {0:8.0f}".format(len(datadf['Age'])))`
	`# Minimum and maximum of the Age variable`
	`print("Minimum: {0:8.3f} Maximum: {1:8.3f}".format(datadf['Age'].`
	`min(), datadf['Age'].max()))`
	`# Mean of the Age variable`
	`print("Mean: {0:8.3f}".format(datadf['Age'].mean()))`
	`# Variance of the Age variable`
	`print("Variance: {0:8.3f}".format(datadf['Age'].var()))`
	`# Standard deviation of the Age variable`
	`print("Standard Deviation: {0:8.3f}".format(datadf['Age'].std()))`
	`##Publications`
	`# To write the name of the output list`
	`print("\nPublications: \n")`
	`# Dimension of the Publications variable`
	`print("Number of elements: {0:8.0f}".format(len(datadf['Publicatio`
	`ns'])))`
	`# Minimum and maximum of the Publications variable`
	`print("Minimum: {0:8.3f} Maximum: {1:8.3f}".`
	`format(datadf['Publications'].min(), datadf['Publications'].max()))`
	`# Mean of the Publications variable`
	`print("Mean: {0:8.3f}".format(datadf['Publications'].mean()))`
	`# Variance of the Publications variable`
	`print("Variance: {0:8.3f}".format(datadf['Publications'].var()))`
	`# Standard deviation of the Publications variable`
	`print("Standard Deviation: {0:8.3f}".format(datadf['Publications'].`
	`std()))`

Outputs	`Age Variable:`
	`Number of elements: 200`
	`Minimum: 24.000 Maximum: 52.000`
	`Mean: 37.056`
	`Variance: 31.779`
	`Standard Deviation: 5.637`
	`Publications:`
	`Number of elements: 200`
	`Minimum: 11.000 Maximum: 70.000`
	`Mean: 29.650`
	`Variance: 59.957`
	`Standard Deviation: 7.743`

With the previous outputs, it is possible to conclude that the variable Age has two missing values (NA's). For the valid values, the age of researchers varies between 24 and 52 years old. The mean (37.06 years) is quite close to the median (37 years), which suggests the non-existence of outliers. The mean is a measure greatly influenced by “large” or “small” values even if these values arise in small number in the sample. When these values (outliers) exist, the mean assumes very different/distant values from the median. This is not true in the case of Age variable.

The first and third quartiles are 33 and 41 years old, respectively. This means that 50% (half) of researchers present in the sample have ages between 33 and 41 years old.

The standard deviation is 5.64 years, that is, on average, the age of researchers varies about six years of the mean (37.06) of their ages.

Regarding the Publications variable, there are no missing values. The number of publications ranges from 11 to 70. It is also possible to check that the mean is 29.65, and the median is 31 publications per researcher. The mean and median are farther apart from each other than in the previous case (Age variable). This suggests the possibility of outliers or suspected outliers existence.

The first and third quartiles are 25 publications and 35, respectively, i.e., half of the researchers have between 25 and 35 publications.

The standard deviation of this variable is approximately 7.74 publications.

Continuous Numerical Variables

A continuous numerical variable can take any numeric value within a specified interval. If the 200 researchers in this analysis were asked to indicate their height, the values should vary a lot. In this case, it is common to proceed with the creation of a set of intervals to group some values. The reader needs to define the size of these ranges. For example, if 500 height records are varying between 1.51m and 1.70m, the amplitude of each range should be small. Otherwise, all values fall into the same interval. If there are 500 salary records between € 1,000 and € 10,000, the amplitude of the interval should be, at least, 1000 or 2000 units.

To show how to work with continuous variables, the Publications variable will be regarded. This variable has been analyzed before. However, the frequencies of each number of publications are unknown. Thus, the frequency analysis (already presented for discrete variables) is provided in Table 13 and Table 14.

As it is possible to observe, the number of publications per researcher varies widely. Therefore, two suggestions of the division with intervals and corresponding frequencies are presented in Table 15 and Table 16, respectively.

Table 13. R language: frequency analysis

	In R

Code	### Frequency analysis of Publications variable
	# # Count of each factor level of the Publications variable
	Freq.Publications<-sort(as.numeric(table(data.df[,"Publications"])),
	decreasing=TRUE)
	# Cumulative frequencies of the "Freq.Publications" object
	CFreq.Freq. Publications <- cumsum(Freq. Publications)
	# Relative frequencies each factor level of the Publications variable
	Rel.Freq <- as.numeric(prop.table(Freq. Publications))
	# Data frame with the realized analysis
	Freqs. Publications <- data.frame(Publications = levels(factor(data.
	df[,"Publications "])), Frequency = Freq. Publications, Cumulative.
	Frequency = CFreq.Freq. Publications, Relative.Frequency = Rel.Freq)

Output	Publications Frequency Cumulative.Frequency Relative.Frequency 1 11 15 15 0.075 2 12 14 29 0.070 3 13 13 42 0.065 4 15 13 55 0.065 5 16 12 67 0.060 6 17 12 79 0.060 7 18 10 89 0.050 8 19 10 99 0.050 9 21 9 108 0.045 10 22 9 117 0.045 11 23 8 125 0.040 12 24 7 132 0.035 13 25 7 139 0.035 14 26 7 146 0.035 15 27 6 152 0.030 16 28 6 158 0.030 17 29 5 163 0.025 18 30 4 167 0.020 19 31 4 171 0.020 20 32 4 175 0.020 21 33 4 179 0.020 22 34 3 182 0.015 23 35 2 184 0.010 24 36 2 186 0.010 25 37 2 188 0.010 26 38 2 190 0.010 27 39 2 192 0.010 28 40 2 194 0.010 29 41 2 196 0.010 30 42 1 197 0.005 31 44 1 198 0.005 32 45 1 199 0.005 33 70 1 200 0.005

In R, in the first case a), 11 intervals are defined. In this case, the amplitude of each interval is fixed. The reader needs only to indicate the number of ranges that he/she wants to consider. To point out that symbol "(" means interval opened, and "]" means range closed. The intervals $\left[11,16\right]$, $\left]16,22\right]$,

Table 14. Python language: frequency analysis

	In Python

Code	`### Frequency analysis of Publications variable`
	`# Filtering Publications data`
	`pubs_datadf = datadf['Publications']`
	`# Group by publications`
	`pubs_datadf = pd.DataFrame(pubs_datadf.value_counts(sort=True))`
	`# Create new column with cumulative sum`
	`pubs_datadf['cum_sum'] = pubs_datadf['Publications'].cumsum()`
	`# Create new column with relative frequency`
	`pubs_datadf['cum_perc'] = 100*pubs_datadf['cum_sum']/pubs_`
	`datadf['Publications'].sum()`
	`pubs_datadf`

Output	`Out[2]:`
	`Publications  cum_sum  cum_perc`
	`31            15 15    7.5`
	`33            14 29    14.5`
	`29            13 42    21.0`
	`25            13 55    27.5`
	`26            12 67    33.5`
	`36            12 79    39.5`
	`39            10 89    44.5`
	`34            10 99    49.5`
	`35            9 108    54.0`
	`32            9 117    58.5`
	`21            8 125    62.5`
	`22            7 132    66.0`
	`24            7 139    69.5`
	`28            7 146    73.0`
	`30            6 152    76.0`
	`38            6 158    79.0`
	`18            5 163    81.5`
	`37            4 167    83.5`
	`40            4 171    85.5`
	`16            4 175    87.5`
	`15            4 179    89.5`
	`19            3 182    91.0`
	`12            2 184    92.0`
	`27            2 186    93.0`
	`41            2 188    94.0`
	`23            2 190    95.0`
	`42            2 192    96.0`
	`44            2 194    97.0`
	`17            2 196    98.0`
	`13            1 197    98.5`
	`70            1 198    99.0`
	`45            1 199    99.5`
	`11            1 200    100.0`

$]22,27]$, $]27,32]$, $]32,38]$, $]38,43]$, $]43,49]$, $]49,54]$, $]54,59]$, $]59,65]$, $]65,70]$ could be considered (round to units). However, this is not a good solution because there are many classes with frequencies equal to one. Thus, some intervals, which seem to make more sense, must be considered.

Table 15. R language: two suggestions of the division with intervals and corresponding frequencies

	In R

Code	### Division at intervals and corresponding frequencies
	a) # Division of the interval in 11 equal parts, and the interval closed on right
	classIntervals(data.df[,”Publications”], n=11, style = “equal”, rtimes = 3,intervalClosure = c(“right”), dataPrecision = NULL)
	
	b) # Division at fixed intervals, i.e., the fixedBreaks indicates the limits of the intervals. The intervals should be closed on right.
	classIntervals(data.df[,”Publications”], n=11, style = “fixed”, fixedBreaks=c(10, 20, 30, 40,70), rtimes = 3,intervalClosure = c(“right”), dataPrecision = NULL)

Outputs	a) style: equal one of 64,512,240 possible partitions of this variable into 11 classes [11,16.36364] (16.36364,21.72727] (21.72727,27.09091] (27.09091,32.45455] (32.45455,37.81818] 12 18 43 50 49 (37.81818,43.18182] (43.18182,48.54545] (48.54545,53.90909] (53.90909,59.27273] (59.27273,64.63636] 24 3 0 0 0 (64.63636,70] 1 b) style: fixed one of 4,960 possible partitions of this variable into 4 classes [10,20] (20,30] (30,40] (40,70] 22 77 93 8

Table 16. Python language: two suggestions of the division with intervals and corresponding frequencies

	In Python

Code	### Division at intervals and corresponding frequencies
	a) # Division of the interval in 11 equal parts, and the interval closed on right
	table = np.histogram(datadf[‘Publications’], bins=11, range=(0, 70))
	print(table)
	
	b) # Division at fixed intervals, i.e., the buckets object indicates the limits of the intervals. The intervals should be closed on right.
	buckets = [0,10, 20, 30, 40,70]
	table = np.histogram(datadf[‘Publications’], bins=buckets)
	print(table)

Outputs	a) (array([0, 3, 19, 37, 55, 64, 20, 1, 0, 0, 1], dtype=int64), array([0.,6.36363636,12.72727273,19.09090909,25.45454545, 31.81818182,38.18181818,44.54545455,50.90909091,57.27272727,63.63636364,70.])) b) array([0, 22, 71, 95, 12], dtype=int64), array([0, 10, 20, 30, 40, 70]))

In the second case b), four intervals are defined. Although it appears that the intervals have different dimensions, it is contemplated that the ends (first and last) include the corresponding infinities, that is, the intervals are $]-\infty,20]$, $]20,30]$, $]30,40]$, $]40,+\infty[$.

Similar to R, in Python, in the first case a), 11 intervals with equal dimension are considered. The first array gives the frequencies of publications in each interval. The defined intervals (rounded to units) are $[0,6]$, $]6,13]$, $]13,19]$, $]19,25]$, $]25,32]$, $]32,38]$, $]38,45]$, $]45,51]$, $]51,57]$, $]57,64]$, $]64,70]$. As there are many classes with frequencies equal to one, the better solution is to set the intervals (presented in b)) manually.

Graphical Representation

To summarize data in a visual way, charts and/or graphs are a good option. Depending on the data type, different graphs should be used. We will explain with some examples.

Pie Chart

The pie chart to represent nominal variables graphically. The data in the study has some variables of this type, to remember, Gender, Python_user, R_user, and Tasks. We provide the pie charts of some of these variables. For Gender, see Table 17 and Table 18.

In R, the more traditional version of a pie chart is shown in output a). Note that the names of each slice should be expressly mentioned. Otherwise, an image without a caption is displayed.

Some packages have been created to make better illustrations of this type of graph. This is the case of package *plotrix* that allows doing 3D pie charts, as shown in output b).

As it is possible to visualize, the size of each slice is proportional to the length of each factor level of the variable. The biggest slice represents the number of male researchers, and the smaller shows the number of female researchers. Thus, the graph indicates that there are 113 males (56.5%) and 87 females (43.5%).

Similarly to R, in Python, the specification of the legend is also required. Also, the pie chart is oval by default. To change it, just indicate that both axes have equal scales. This is a condition that provides a circle pie chart.

Table 17. R language: gender pie chart

<table>
<tr><th></th><th>In R</th></tr>
<tr><td>Code</td><td>### Gender Pie Chart
# Frequency of the Gender variable
mytable <- table(data.df[,"Gender"])
# Labels of each pie slice. Past gender classes with their frequencies
lbls <- paste(names(mytable), "\n", mytable, sep="")
a)
Graph with labels lbls and a name of the pie chart
pie(mytable, labels = lbls,
main="Pie Chart of Gender Variable\n (with sample sizes)")

OR b)
# A 3D Graph with labels lbls, spacing between the slices (input *explode*) and a name of the pie chart
library(plotrix)
pie3D(mytable, labels = lbls, explode=0.1,
main="Pie Chart of Gender Variable\n (with sample sizes)")</td></tr>
<tr><td>Outputs</td><td>a) Pie chart of Gender variable in R:
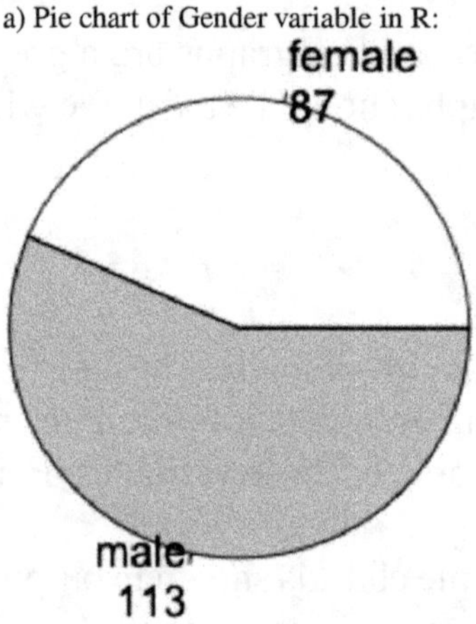

b) 3D Pie chart of Gender variable in R:
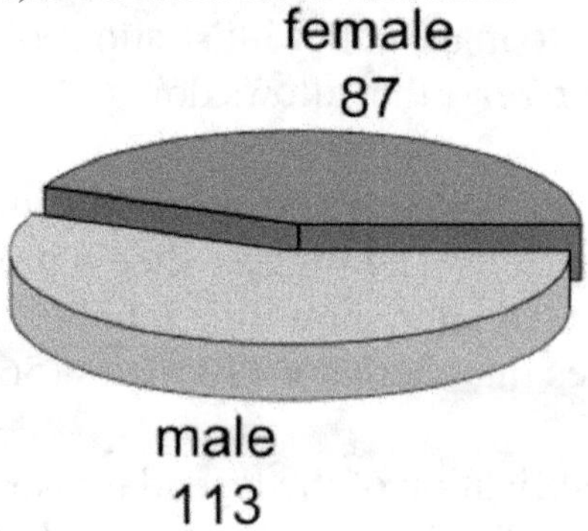
</td></tr>
</table>

Table 18. Python language: gender pie chart

	In Python

Code	```### Pie Chart Gender```
	 ```# Import packages matplotlib.pyplot and pandas```
	 ```import matplotlib.pyplot as plt```
	 ```import pandas as pd```
	 ```# Pie chart labels```
	 ```label_list = datadf['Gender'].value_counts(sort=False).index```
	 ```# Plot pie chart axis```
	 ```plt.axis("equal")```
	 ```# The pie chart is oval by default. To make it a circle use pyplot.```
	 ```axis("equal")```
	 ```# To show the percentage of each pie slice, pass an output format to```
	 ```the autopct parameter (rounding to 1 decimal place).```
	 ```plt.pie(datadf['Gender'].value_counts(sort=False),labels=label_```
	 ```list,autopct="%1.1f%%")```
	 ```plt.title("Researchers Gender")```
	 ```plt.show()```

Output	Pie chart of Gender variable in Python: 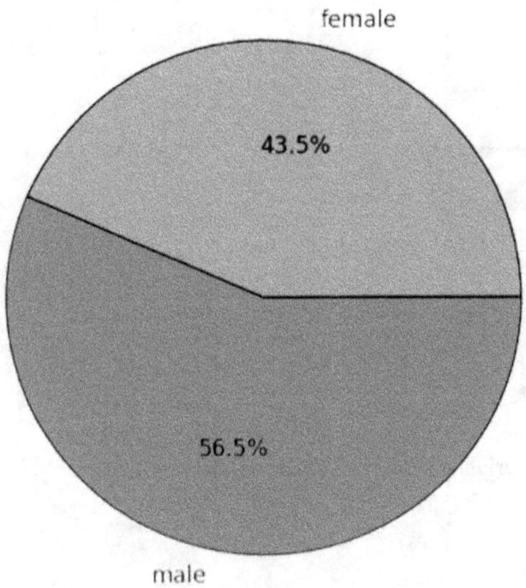

The values of frequencies (and percentage), as expected, coincide with the relative frequencies analyzed at the beginning of this chapter.

Regarding the Python_user variable, similar pie charts will be done in Table 19 and Table 20.

Regarding the Python_users variable, the R pie charts above show that there are 107 users ($\approx$54%) and 92 non-users ($\approx$46%). We should point out that missing values are not represented. They are just eliminated.

Table 19. R language: Python_user variable pie chart

	In R

Code	### Python_user
	a) # Graph with labels lbls and a name of the pie chart
	mytable <- table(data.df[,"Python_user"])
	lbls <- paste(names(mytable), "\n", mytable, sep="")
	pie(mytable, labels = lbls,
	main="Pie Chart of Python users Variable\n (with sample sizes)")
	
	b) # A 3D Graph with labels lbls, spacing between the slices (input *explode*) and a name of the pie chart
	library(plotrix)
	pie3D(mytable, labels = lbls, explode=0.1,
	main="Pie Chart of Python users Variable\n (with sample sizes)")

Outputs	a) Pie chart of Python_user variable in R: 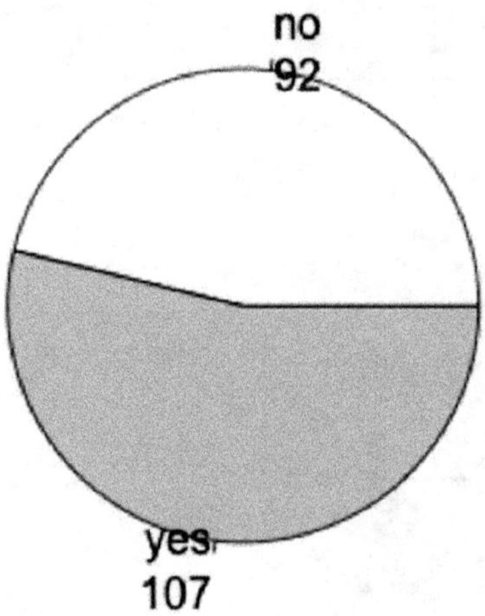 b) 3D Pie chart of Python_user variable in R: 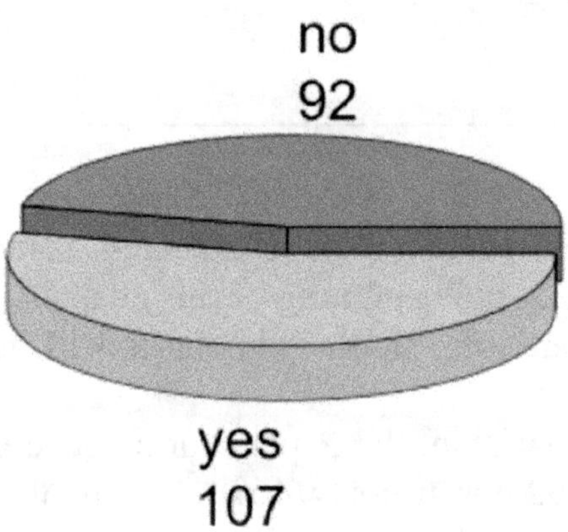

Table 20. Python language: Python_user variable pie chart

<table>
<tr><th></th><th>In Python</th></tr>
<tr><td>Code</td><td><pre><code>### Python_user
# Pie Chart Python_user
import matplotlib.pyplot as plt
import pandas as pd
# Pie chart labels
label_list = datadf['Python_user'].value_counts(sort=False).index
plt.axis("equal") #The pie chart is oval by default. To make it a
circle use pyplot.axis("equal")
To show the percentage of each pie slice, pass an output format
to the autopct parameter
plt.pie(datadf['Python_user'].value_
counts(sort=False),labels=label_list,autopct="%1.1f%%")
plt.title("Researchers Python Users")
plt.show()</code></pre></td></tr>
<tr><td>Output</td><td>Pie chart of Python_user variable in Python:
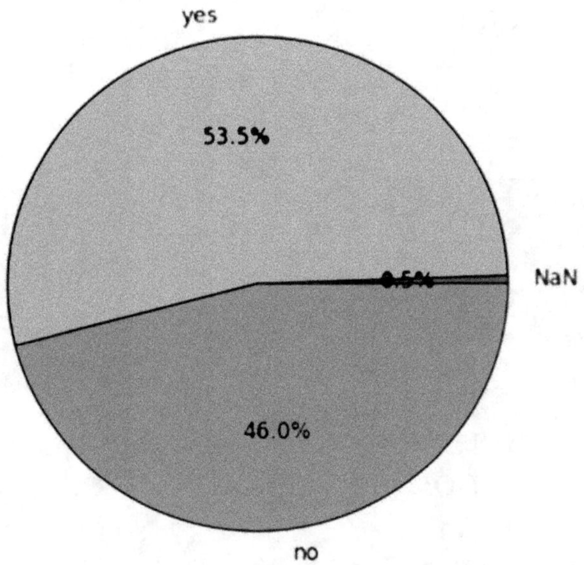
</td></tr>
</table>

In Python, missing values are represented in the pie chart. Thus, the Python output shows that there are 53.5% of users, 46% of non-users and 0.5% of missing values. If the number of absence answers is not of reader's interest, the correspondent users could be deleted, or a condition in the plot should be inserted.

A similar pie chart for the R_users variable is provided in Table 21 and Table 22.

Table 21. R language: R_user variable pie chart

	In R

Code	`### R_user`
	`a) # Graph with labels lbls and a name of the pie chart`
	`mytable <- table(data.df[,"R_user"])`
	`lbls <- paste(names(mytable), "\n", mytable, sep="")`
	`pie(mytable, labels = lbls,`
	`main="Pie Chart of R users Variable\n (with sample sizes)")`
	
	`b) # A 3D Graph with labels lbls, spacing between the slices (input` *`explode`*`) and a name of the pie chart`
	`library(plotrix)`
	`pie3D(mytable, labels = lbls,explode=0.1,`
	`main="Pie Chart of R users Variable\n (with sample sizes)")`

Outputs	a) Pie chart of R_user variable in R: 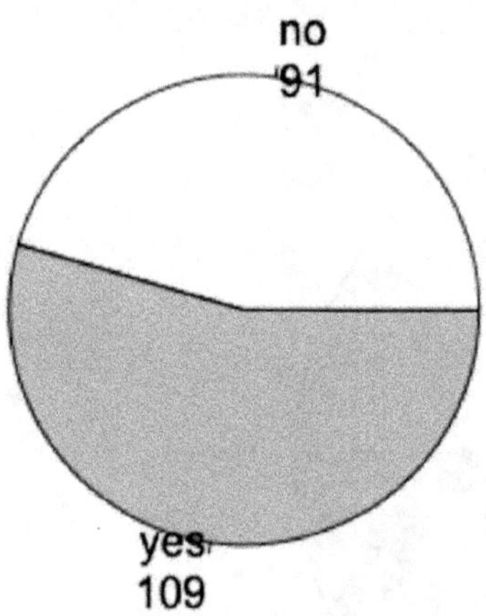 b) 3D Pie chart of R_user variable in R: 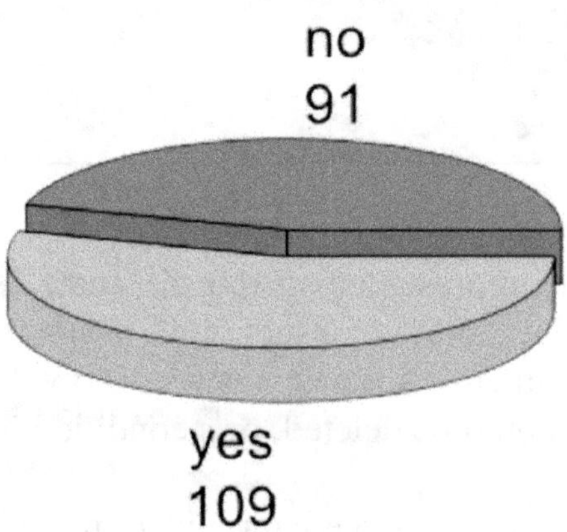

Table 22. Python language: R_user variable pie chart

	In Python

Code	`### R_user`
	`# Pie Chart R_user`
	`import matplotlib.pyplot as plt`
	`import pandas as pd`
	`# Pie chart labels`
	`label_list = datadf['R_user'].value_counts(sort=False).index`
	`plt.axis("equal") #The pie chart is oval by default. To make it a circle use pyplot.axis("equal")`
	`# To show the percentage of each pie slice, pass an output format to the autopct parameter`
	`plt.pie(datadf['R_user'].value_counts(sort=False),labels=label_list,autopct="%1.1f%%")`
	`plt.title("Researchers R Users")`
	`plt.show()`

Output	Pie chart of R_user variable in Python: 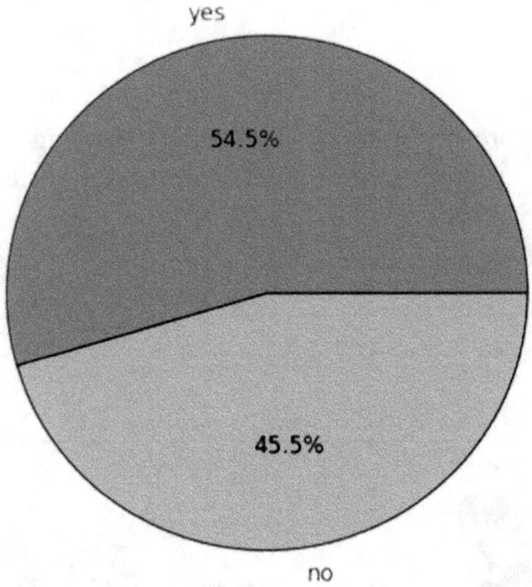

The R pie charts in Tables 21 and 22 show that there are 109 R users (54.5%) and 91 non-users (45.5%). In the case of this variable (R_users), there are no missing values. The results presented in the pie chart correspond to the entire sample. As expected, the values are the same as initially obtained.

Bar Graph

Similar to pie charts, the bar graphs are very useful in case of discrete variables, namely nominal variables. In the case of Tasks variable, either a pie chart or a bar graph is suitable. Examples of bar graphs are shown in Table 23 and Table 24.

In the previous outputs, a bar graph of the Tasks variable is represented. The x-axis represents the different factor levels of the Tasks variable. In the y-axis, the frequencies of each factor level. The code to create this bar graph specifies the length of the y-axis and the maximum is 78. Thus, the largest bar represents the number of Ph.D. Students. The Ph.D. Supervisors are shown in the shortest bar, in which the maximum frequency is near 60. Finally, the number of Postdoctoral researchers is near 70.

Boxplots

One of the best ways to represent numerical variables is using a boxplot. This type of graph allows creating a general idea about the variable since

Table 23. R language: "Tasks" bar graph

	In R

Code	`### Tasks Bar Graph`
	`# Graph with different colors in the bars`
	`barplot(table(data.df[,"Tasks"]), col=c(1,2,3), ylim = c(0, 80))`

Output	Bar graph of Tasks variable in R: 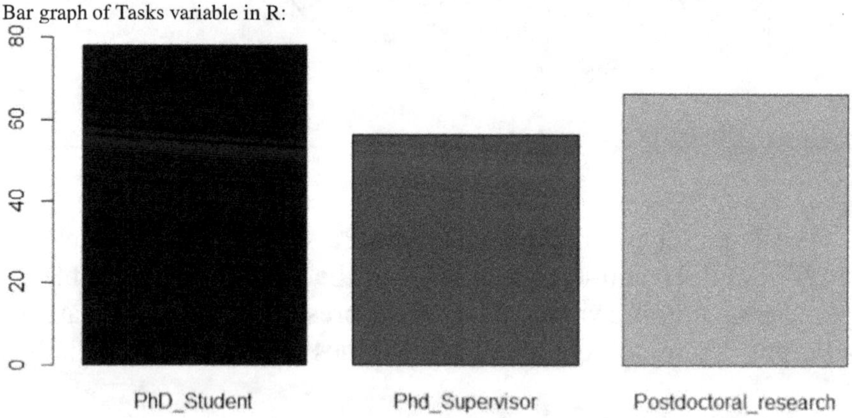

Table 24. Python language: "Tasks" bar graph

	In Python

Code	`### Tasks Bar Graph`
	`# Grouped sum of Tasks`
	`var = datadf['Tasks'].value_counts(sort=False)`
	`# Tasks Bar Graph`
	`plt.figure()`
	`# Setting y-axis label`
	`plt.ylabel('Number of Reseachers')`
	`# Setting graph title`
	`plt.title("Counting Reseacher's Tasks")`
	`# Trigger Bar Graph`
	`var.plot.bar()`
	`# Show Bar Graph`
	`plt.show()`

Output	Bar graph of Tasks variable in Python: 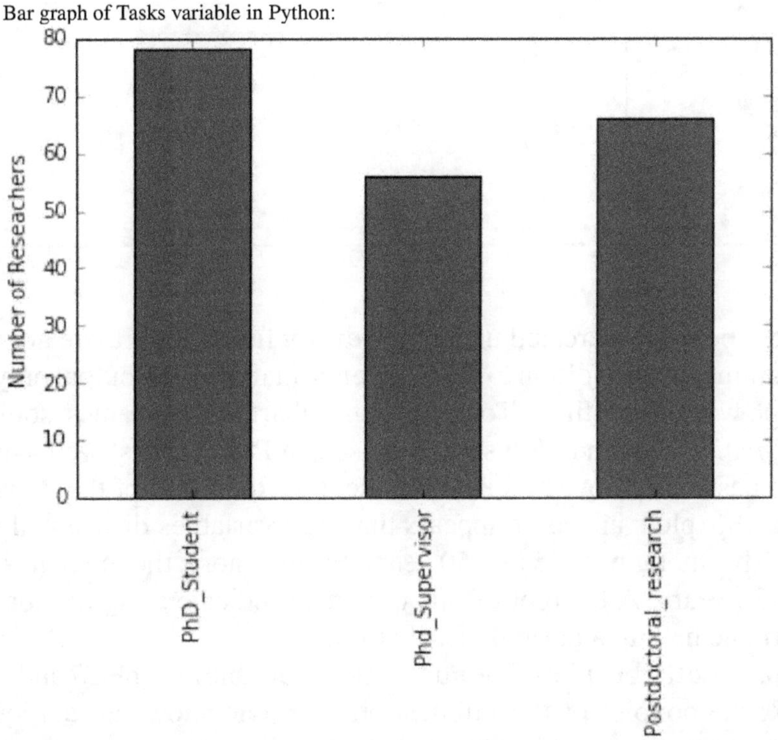

some critical values are given. Median, quartiles, maximum, minimum and outliers (if it exists) can be observed. We provide the boxplot of Age and Publications variables in Table 25 and Table 26.

Table 25. R language: boxplot of Age variable

<table>
<tr><th></th><th>In R</th></tr>
<tr><td>Code</td><td>

```
### Boxplot of Age variable
# Boxplot with title and name of the y-axis
boxplot(data.df[,"Age"],data=data.df, main="Scientific Researchers
Data",xlab="", ylab="Age of Researchers")
```

</td></tr>
<tr><td>Output</td><td>Boxplot of Age variable in R:

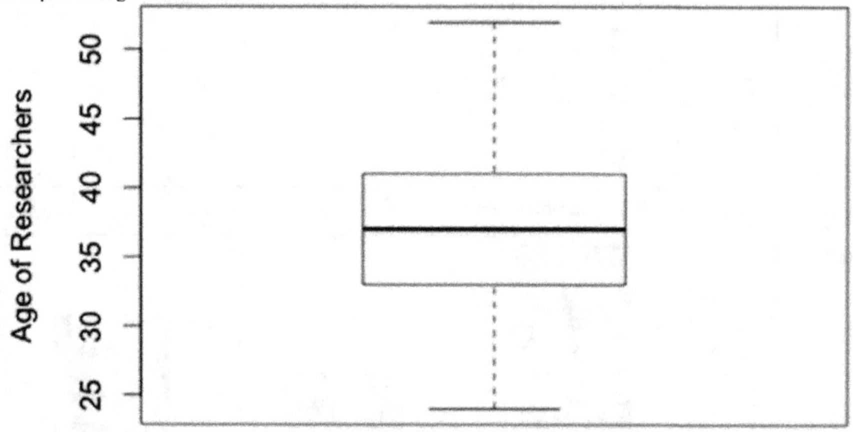

</td></tr>
</table>

In R, a boxplot is created using the *boxplot* function. If there are no other input parameters in addition to the data being analyzed, the output only returns the graph without subtitles. To be given the chart or axes names, some inputs such as *main*, *xlab* and *ylab* should be set. In Python, first we have to take NaN values and remove or substitute them for the mean of the variable.

In the boxplots above, it appears that Age variables distributed more or less evenly among near 25 and 50 years. Furthermore, the median is slightly over to 35 years. A boxplot of this type may indicate a high probability of this variable having a normal distribution.

We provide the boxplot of the Publications variable in Table 27 and Table 28.

In R, the boxplot of the Publications variable shows an outlier, represented by a small circle. This means that there is only one person with a high number of publications when compared with the others. Also, the average of publications is around 30.

Table 26. Python language: boxplot of Age variable

<table>
<tr><th></th><th>In Python</th></tr>
<tr><td>Code</td><td>

```
### Boxplot of Age variable
# First we have to take NaN values and substitute them for the mean of
the variable
datadf['Age']=datadf['Age'].replace('nan',datadf['Age'].mean())
# Importing modules
import matplotlib.pyplot as plt
import pandas as pd
# Setting Figure
fig=plt.figure()
ax = fig.add_subplot(1,1,1)
# Triggering Boxplot Chart
ax.boxplot(datadf['Age'],showfliers=True, flierprops =
dict(marker='o', markerfacecolor='green', markersize=12,linestyle='no
ne'))
# Setting Plot Title
plt.title('Age Boxplot')
# Show Plot
plt.show()
```

</td></tr>
<tr><td>Output</td><td>Boxplot of Age variable in Python:

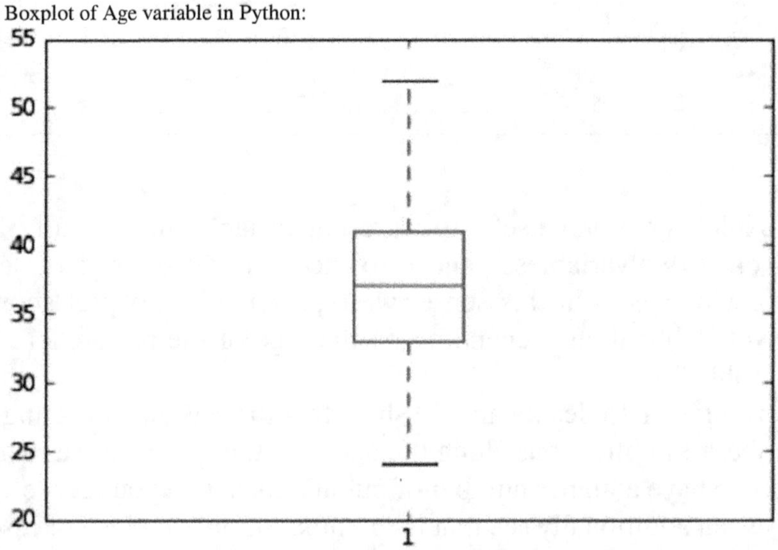

</td></tr>
</table>

The distribution of this variable seems to be quite irregular.

In Python, by default, the outliers are removed. If the analysts need to visualize them, the command *showfliers=True* and some particularities of the *flierprops* command should be included.

The Python's boxplot above, similar to R, shows the variation of the Publications variable, such as median, maximum, minimum and quartiles.

Table 27. R language: boxplot of Publications variable

	In R

Code	`### Boxplot of Publications variable`
	`# Boxplot with title and name of the y-axis`
	`boxplot(data.df[,"Publications"],data=data.df, main="Scientific Researchers Data",xlab="", ylab="Pubications of Researchers")`

Output	Boxplot of Publications variable in R: 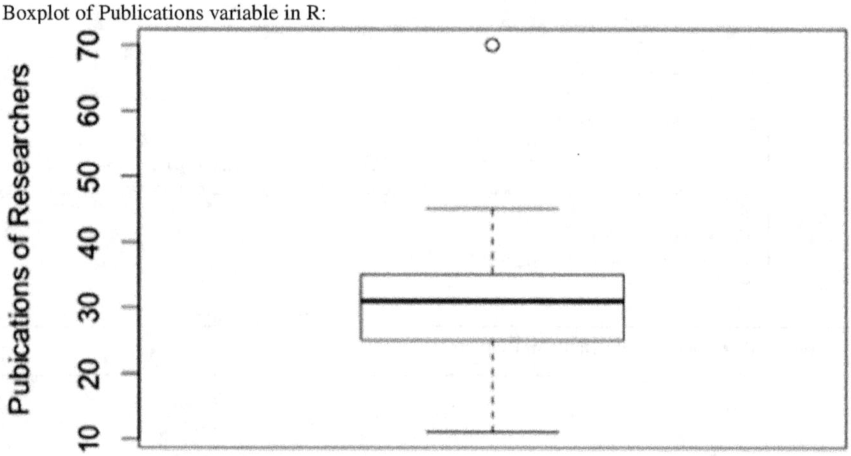

A boxplot is not only useful to represent variables individually. It is possible to cross two variables x_1 and x_2 to check the behavior of x_2, depending on x_1, or vice versa. In this sense, we represented a boxplot showing the number of Publications depending on the Age of the researchers in Table 29 and Table 30.

The figures in Tables 29 and 30 show that there is an increasing trend in the number of publications along the age, i.e., the senior researchers have a tendency to have a higher number of publications than younger researchers. Also, we can additionally see that the number of publications of researchers between 30 and 40 years old, and more than 50 years old, varies more than in other age intervals.

Histogram

As the intervals of the number of Publications are continuous, a histogram should be used as a visual representation. See Table 31 and 32. The big difference between the histogram and the bar graph essentially is in the x-axis.

Table 28. Python language: boxplot of Publications variables

<table>
<tr><th></th><th>In Python</th></tr>
<tr><td>Code</td><td>

```
### Boxplot of Publications variable
# Import packages matplotlib.pyplot and pandas
import matplotlib.pyplot as plt
import pandas as pd
# Plot boxplot and create one or more subplots using add_subplot,
because you can't create blank figure
fig=plt.figure()
ax = fig.add_subplot(1,1,1)
# Triggering Boxplot Chart
ax.boxplot(datadf['Publications'],showfliers=True, flierprops =
dict(marker='o', markerfacecolor='green', markersize=12,linestyle='no
ne'))
# Boxplot Title
plt.title('Publications Boxplot')
# Show Boxplot
plt.show()
```

</td></tr>
<tr><td>Output</td><td>Boxplot of Publications variable in Python:
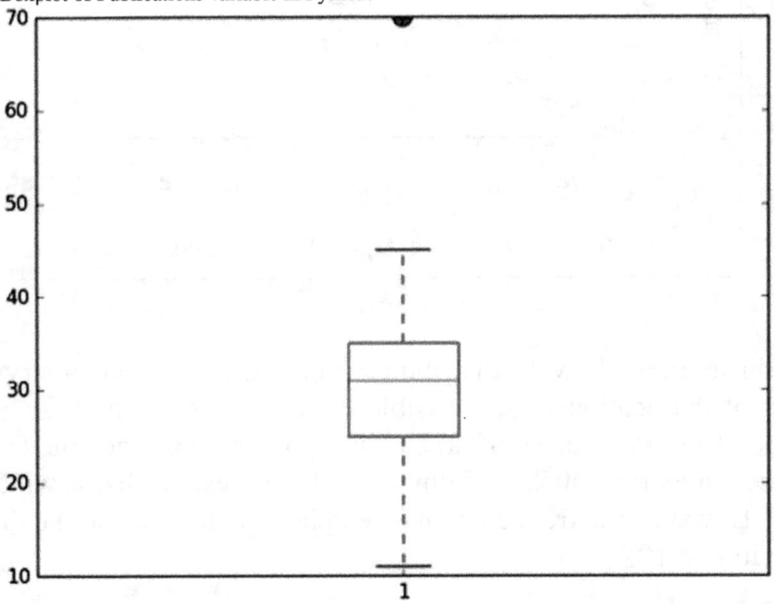
</td></tr>
</table>

Instead, on the bar graph, the frequencies are represented in separate bars, the histogram has continuous bars. In both programming languages, some particularities can be specified, namely the axis labels and the main label of the histogram. Otherwise, the output only returns the picture without labels.

Table 29. R language: boxplot showing the number of Publications depending on the Age of the researchers

	In R

Code	`### Boxplot of Publications vs. Age`
	`boxplot(Publications~Age,data=data.df, main="Scientific Researchers Data",`
	`xlab="Age of Researchers", ylab="Number of Publications")`

Output	Boxplot of the Publications depending on the Age in R: 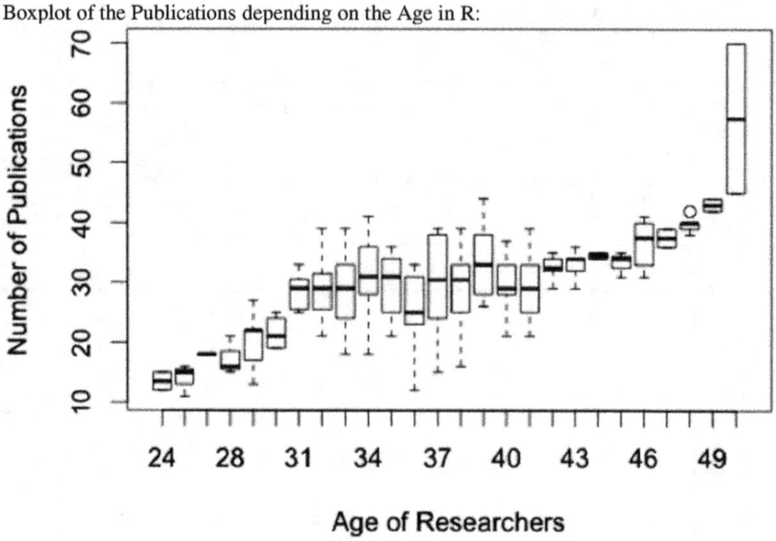

The histograms show the distribution of frequencies along intervals of the number of Publications. It is possible to verify that the highest frequency corresponds to the interval of 30 and 40 publications. Since the last bar includes researchers publishing from 40 to 70 articles, this bar is wider (in R's output). However, the frequency of researchers with 40 or more publications is only around 10.

How to Present Results?

Regardless the programming language, the presentation of results is crucial. Mostly, the outputs provide a lot of information. Some of this information is not of the reader's interest, or it is not necessary to be demonstrated. On the other hand, depending on how this information is synthesized, it can become easier or more difficult to read.

Table 30. Python language: boxplot showing the number of Publications depending on the Age of the researchers

<table>
<tr><th></th><th>In Python</th></tr>
<tr><td>Code</td><td><pre>### Boxplot of Publications vs. Age
# Import pandas package
import pandas as pd
# First we have to take NaN values and substitute them for the mean of
the variable
datadf['Age']=datadf['Age'].replace('nan',datadf['Age'].mean())
# Get only Publications and Age variables
new_datadf = datadf.ix[:,['Age','Publications']]
# Boxplot
new_datadf.boxplot(by='Age',rot = 90)</pre></td></tr>
<tr><td>Output</td><td>Boxplot of the Publications depending on the Age in Python:

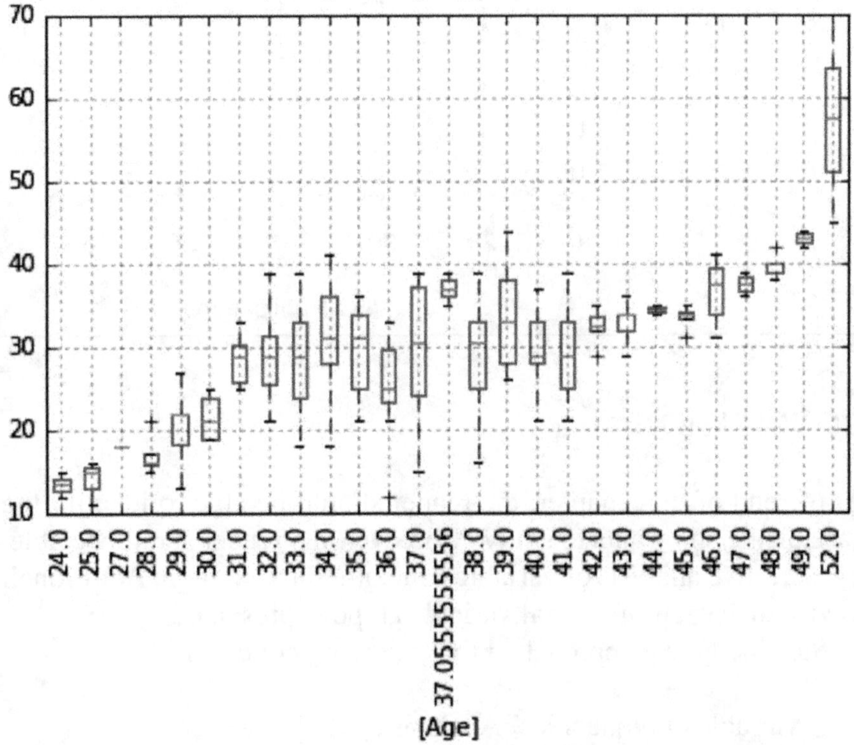

</td></tr>
</table>

Thus, in this sense, a presentation of the analysis made throughout this chapter is proposed. Note that, graphically, the user only presents the chart that considers more appropriate. The remaining analysis can be systematized like in Table 33.

Table 31. R language: histogram of the number of Publications (defined by classes)

	In R
Code	`### Histogram of the number of Publications (defined by classes)` `hist(data.df[,"Publications"],breaks=c(10, 20, 30, 40,70), right = TRUE, freq=TRUE, ylab = "Frequencies", xlab="# of Publications", main="Histogram for Publications",include.lowest=TRUE)`
Output	Histogram of the Publications in R: 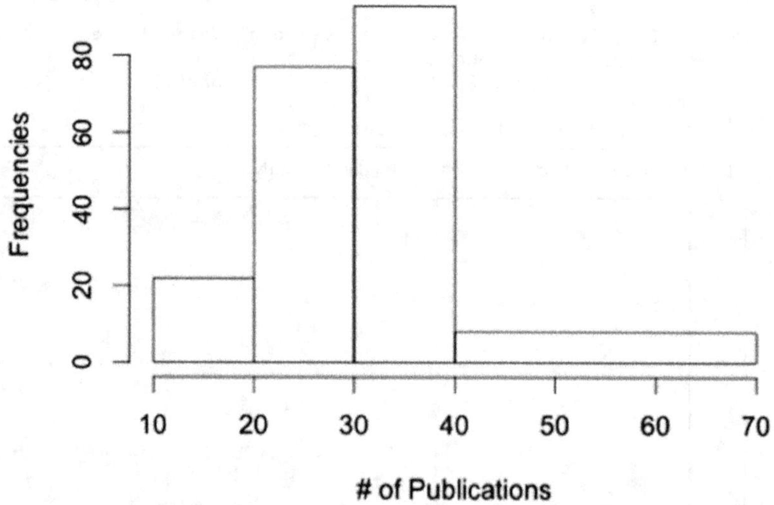

CONCLUSION

At the end of this chapter, the reader should be able to identify the type of variable he/she wants to study. Moreover, the reader should be able to make a descriptive analysis of variables, in both R and Python. Additionally, a way to visualize each of several variables type is presented.

Succinctly, we approached the following concepts:

- Variables Frequencies' Analysis.
- Measures of Central Tendency and Dispersion:
  - Median,
  - Mean,
  - Standard Deviation,
  - Max,
  - Min,
  - Quartiles.

Table 32. Python language: histogram of the number of Publications

<table>
<tr><th></th><th>In Python</th></tr>
<tr><td>Code</td><td>

```
### Histogram of the number of Publications
# Publications Histogram
fig=plt.figure() #Plots in matplotlib reside within a figure object,
use plt.figure to create new figure
# Create one or more subplots using add_subplot, because you can't
create blank figure
ax = fig.add_subplot(1,1,1)
# Variable
ax.hist(datadf['Publications'],facecolor='green')# Here it is possible
to change the number of bins
# Limits of x axis
ax.set_xlim(0, 70)
# Limits of y axis
ax.set_ylim(0,80)
# Set grid on
ax.grid(True)
# Labels and Title
plt.title('Publications distribution')
plt.xlabel('Publications')
plt.ylabel('#Reseachers')
# Finally, show plot
plt.show()
```

</td></tr>
<tr><td>Output</td><td>Histogram of the Publications in Python:
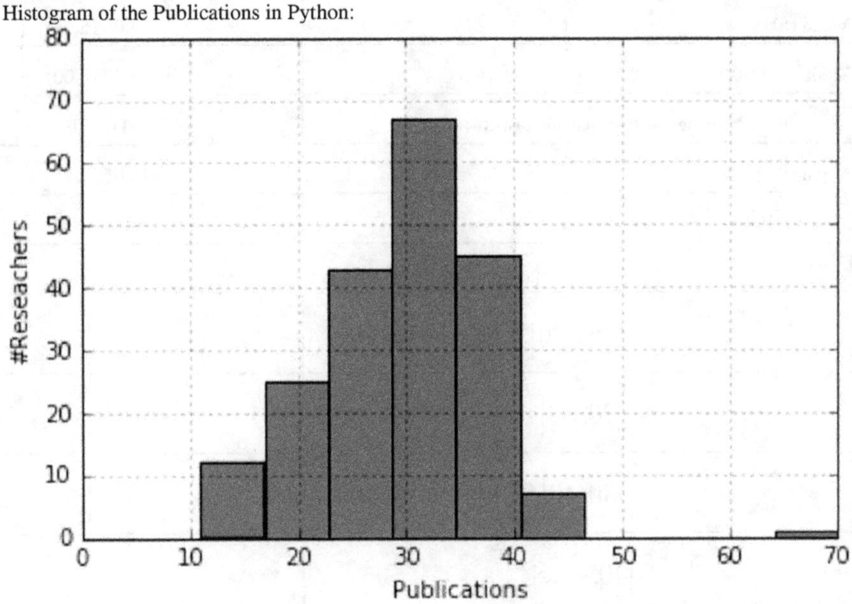

</td></tr>
</table>

Table 33. How to present results

	N	%
Gender		
Male	113	56.5
Female	87	43.5
Users		
Python users*	107	53.7
R users	109	54.5
Tasks		
PhD Student	78	39
PhD Supervisor	56	28
Postdoctoral Research	66	33
Age (min-max)**	**24 – 52**	
Mean (SD)	37.06 (5.64)	
Median (IQR)	37.00 (8.00)	
Number of Publications (min-max)	**11 – 70**	
Mean (SD)	29.65 (7.74)	
Median (IQR)	31.00 (10.00)	
Number of Publications (Intervals)		
$]-\infty, 20]$	22	11.0
$]20, 30]$	77	38.5
$]30, 40]$	93	46.5
$]40, +\infty[$	8	4.0

* N=199; **N=198; IQR – Interquartile Range; SD – Standard Deviation.

- Outlier and missing values.
- Graphical representation of variables:
    - Pie Chart,
    - Bar Graph,
    - Boxplot,
    - Histogram.

Chapter 5
Statistical Inference

INTRODUCTION

Statistical inference allows drawing conclusions from data that might not be immediately apparent. These analyses use a random sample of data taken from a population to describe and make inferences about the population. Inferential statistics are valuable when it is not convenient or possible to examine each member of an entire population.

In this chapter, some concepts like ANOVA, Student's t-test, Chi-Square test, Mann-Whitney test, Kruskal-Wallis test, etc., will be presented.

R VS. PYTHON

To make an inferential statistical analysis, the first step is checking the normality of the numerical variables. Depending on the normality or non-normality of the variables, parametric or non-parametric tests, respectively, should be used. In the following sections, we will present some suggestions to do inferential statistical analysis, in R and Python.

Normality Tests

An assessment of the normality of data is a prerequisite for many statistical tests because normal data is an underlying assumption in parametric tests.

DOI: 10.4018/978-1-68318-016-6.ch005

There are two primary methods of assessing normality: numerically or graphically, relying on statistical tests or visual inspection, respectively.

Statistical tests have the advantage of making an objective judgment of normality, but there are disadvantaged because, sometimes, they are not sensitive enough at low sample sizes or may be overly sensitive to large sample sizes. As such, some statisticians prefer to use their experience to make a subjective judgment about the data from plots/graphs. Graphical interpretation has the advantage of allowing an experienced researcher to assess normality in situations when numerical tests might be over or under sensitive, but graphical methods do lack objectivity. If the reader does not have enough experience interpreting normality graphically, it is probably best practice to rely on the numerical methods.

The data used in this book has two numerical variables, the Age, and the Publications. So, the programming of normality tests for these variables will be done numerically and graphically, using R and Python languages.

Numerically

To test the normality numerically, some statistical tests should be used. In a case of small samples (less than 50 records/users), Shapiro-Wilk test is more convenient. On the opposite, if the sample has more than 50 records/users, Kolmogorov-Smirnov test should be used.

Despite Age and Publications variables have more than 50 records, and Kolmogorov-Smirnov is more appropriate, in this book, the normality test with both tests are presented.

The analyses in R and Python are presented in Table 1 and Table 2.

Table 1. R language: Shapiro-Wilk normality test for Age and Publications variables

	In R

Code	`### Shapiro-Wilk normality test for Age and Publications variables`
	`# na.aggregate substitutes NA's by the mean of the variable`
	`shapiro.test(na.aggregate(data.df$Age))`
	`shapiro.test(na.aggregate(data.df$Publications)`

Outputs	`Shapiro-Wilk normality test`
	`data: na.aggregate(data.df$Age)`
	`W = 0.9921, p-value = 0.3523`
	`Shapiro-Wilk normality test`
	`data: data.df$Publications`
	`W = 0.9592, p-value = 1.624e-05`

Table 2. Python language: Shapiro-Wilk normality test for Age and Publications variables

	In Python

Code	### Shapiro-Wilk normality test for Age and Publications variables
	# Import packages
	import scipy
	from scipy import stats
	# Substitution of the NA's by the mean of the Age variable
	datadf['Age']=datadf['Age'].replace('nan',datadf['Age'].mean())
	# Shapiro-Wilk test for Age and Publications variables
	print(stats.shapiro(datadf['Age']))
	print(stats.shapiro(datadf['Publications']))

Outputs	# Shapiro-Wilk test for Age and Publications variables, respectively (0.9920958876609802, 0.3522576689720154) (0.9592043161392212, 1.6238263924606144e-05)

For the outputs shown in Tables 1 and 2, Shapiro-Wilk test allows concluding that, with a 95% of confidence, the null hypothesis is not rejected (p=0.3523>0.05) for the Age variable, i.e., there is no evidence to reject the null hypothesis and it may be considered the existence of normality. For Publications variable, Shapiro-Wilks suggests the rejection of the null hypothesis (p<0.0001<0.05), i.e., this variable does not have a normal distribution (with a significance level of 5%). Hence, based on this test, it may be considered the existence of normality for the Age and the non-normality for the Publications variable (with a significance level of 5%).

In some software, there is a one-sample Kolmogorov-Smirnov test. This test allows verifying if the frequencies of one variable have distribution near the normal. However, in R, we need two distributions of values. Therefore, an *rnorm* function is used to generate a distribution with 200 values and mean of the variable in the study. After making this distribution, the distribution of Age or Publications variables is compared, to verify the equality.

Thus, the hypotheses are:

$\mathbf{H_0}$**:** The variable has normal distribution.
$\mathbf{H_1}$**:** The variable has not normal distribution.

Since with Python, the Kolmogorov-Smirnov test returns some unexpected values, we decided to use the R software in Python, and proceed with this analysis. Here is the Kolmogorov-Smirnov test using both languages in Table 3 and Table 4.

Table 3. R language: Kolmogorov-Smirnov normality test for Age and Publications variables

	In R

Code	`### Kolmogorov-Smirnov normality test for Age and Publications variables`
	`# Generate a rnorm with 200 values in order to compare distributions`
	`ks.test(na.aggregate(data.df$Age), rnorm(200, mean(na.aggregate(data.df$Age)))`
	`ks.test(data.df$Publications, rnorm(200, mean(data.df$Publications)))`

Outputs	`Two-sample Kolmogorov-Smirnov test`
	`data: na.aggregate(data.df$Age) and rnorm(200, mean (na.aggregate(data.df$Age)), sd(na.aggregate(data.df$Age)))`
	`D = 0.13, p-value = 0.06809`
	`alternative hypothesis: two-sided`
	`Two-sample Kolmogorov-Smirnov test`
	`data: data.df$Publications and rnorm(200, mean (data.df$Publications))`
	`D = 0.43, p-value = 2.22e-16`
	`alternative hypothesis: two-sided`

Kolmogorov-Smirnov test (recommended with the data type of this study) in R, suggests that the reader should not reject the null hypothesis ($p=0.06809>0.05$) for the Age variable. The same test suggests rejecting the null hypothesis for the Publications variable ($p<0.0001<0.05$).

Regarding the Kolmogorov-Smirnov test in Python, it is possible to observe the different values of p (when compared with R results). These differences could be explained because an *rnorm* distribution was generated in several moments. Although both tests of the same Age variable are normal, the individual values are different. Furthermore, the same conclusions as in R could be obtained.

As both normality tests suggest the normality of Age and the non-normality of Publications, the following analysis along this chapter, should use parametric tests with Age and non-parametric tests with Publications. See Table 5.

Graphically

Graphically, for testing the normality or non-normality of variables, a QQ-plot could be done. If the representation of the points is near a straight-line $y=x$, the normal distribution of the variable could be assumed. Otherwise, the variable does not have a normal distribution. See Table 6 and Table 7 and the figures contained within.

Table 4. Python language: Kolmogorov-Smirnov normality test for Age and Publications variables

	In Python

Code	<pre>### Kolmogorov-Smirnov normality test for Age and Publications
	variables
	# Import packages
	from pandas import *
	import rpy2 as rpy2
	from rpy2.robjects.packages import importr
	import rpy2.robjects as ro
	import pandas.rpy.common as com
	# Changing R's directory to where the data is
	ro.r('setwd("C:/Users/Rui Sarmento/Documents/Livro Cybertech/
	Dados e Code")')
	# Reading the data with R
	ro.r('data_df <- read.csv("data.csv",sep=";")')
	# Reading the R's package zoo, needed to apply na.aggregate
	function
	ro.r('library(zoo)')
	# Kolmogorov-Smirnov Normal Distribution test for Age and
	Publications variables
	print(ro.r('ks.test(na.aggregate(data_df$Age), rnorm(200, mean
	(na.aggregate(data_df$Age)), sd (na.aggregate(data_df$Age))))'))
	print(ro.r('ks.test(data_df$Publications, rnorm(200, mean(data_
	df$Publications)))'))</pre>

Outputs	<pre># Kolmogorov-Smirnov normality test for Age and Publications
	variables, respectively
	print(ro.r('ks.test(na.aggregate(data_df$Age), rnorm(200, mean(na.
	aggregate(data_df$Age)), sd(na.aggregate(data_df$Age))))'))
	Two-sample Kolmogorov-Smirnov test
	data: na.aggregate(data_df$Age) and rnorm(200, mean(na.
	aggregate(data_df$Age)), sd(na.aggregate(data_df$Age)))
	D = 0.09, p-value = 0.3927
	alternative hypothesis: two-sided
	print(ro.r('ks.test(data_df$Publications, rnorm(200, mean(data_
	df$Publications)))'))
	Two-sample Kolmogorov-Smirnov test
	data: data_df$Publications and rnorm(200, mean(data_df$Publications))
	D = 0.425, p-value = 4.441e-16
	alternative hypothesis: two-sided</pre>

Table 5. To synthesize (with Kolmogorov-Smirnov test)

Variable	p	H_0	Normality	Tests
Age	0.06809 > 0.05	Not rejected	Normal	Parametric
Publications	0.0001 <0.05	Rejected	Non-normal	Non-parametric

Table 6. R language: QQ-plots

<table>
<tr><th></th><th>In R</th></tr>
<tr><td>Code</td><td>### QQ-plots
#Needed package
library(zoo)
#Take NAs out
na.aggregate(data.df$Age)
na.aggregate(data.df$Publications)

a) # Test normality with normal Q-Q Plot for Age variable
qqmath(data.df$Age, distribution = qnorm, type = c("p", "g"),
aspect = "xy", pch = ".", cex = 2, ylab="Age")

b) # Test normality with normal Q-Q Plot for Publications variable
qqmath(data.df$Publications, distribution = qnorm, type = c("p", "g"),
aspect = "xy", pch = ".", cex = 2, ylab="Publications")</td></tr>
<tr><td>Outputs</td><td>a) QQ-plot for Age variable in R:
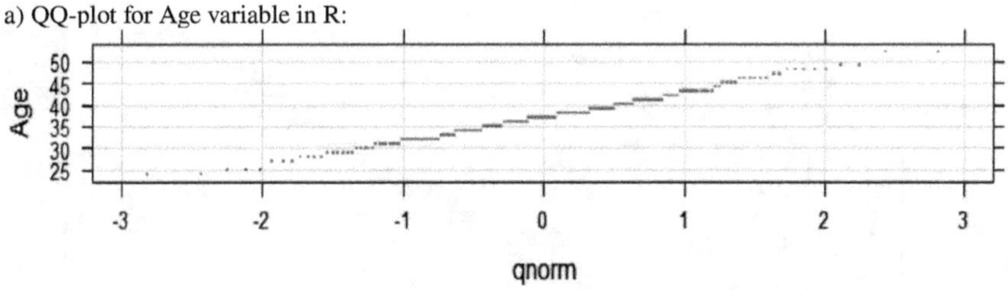

b) QQ-plot for Publications variable in R:
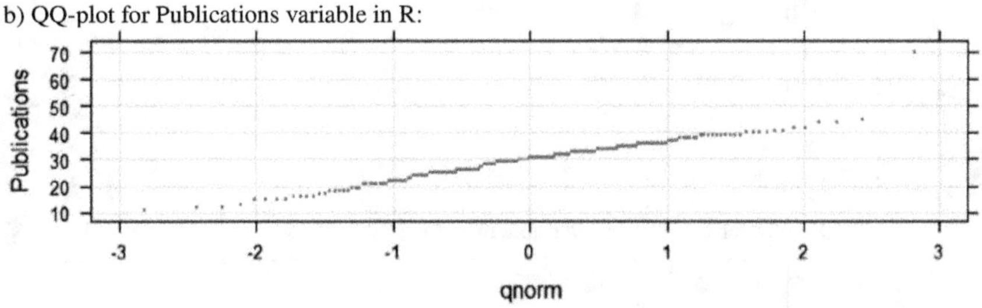
</td></tr>
</table>

Based on QQ-plots, it is possible to verify that the representation of the Age variable has a tendency to be straight-line. Otherwise, for the Publications variable, the line has a small curve at the end, excluding the possibility of a normal distribution. Thus, based on the previous outputs, the normality of the Age and the non-normality of the Publications were assumed.

Table 7. Python language: QQ-plots

<table>
<tr><th></th><th>In Python</th></tr>
<tr><td>Code</td><td>

```
### QQ-plots
# Needed package
import pylab
import scipy.stats as stats

a) # Test normality with normal Q-Q Plot for Age variable
stats.probplot(datadf['Age'], dist="norm", plot=pylab)
pylab.show()

b) # Test normality with normal Q-Q Plot for Publications
variable
stats.probplot(datadf['Publications'], dist="norm", plot=pylab)
pylab.show()
```

</td></tr>
<tr><td>Outputs</td><td>

a) QQ-plot for Age variable in Python:

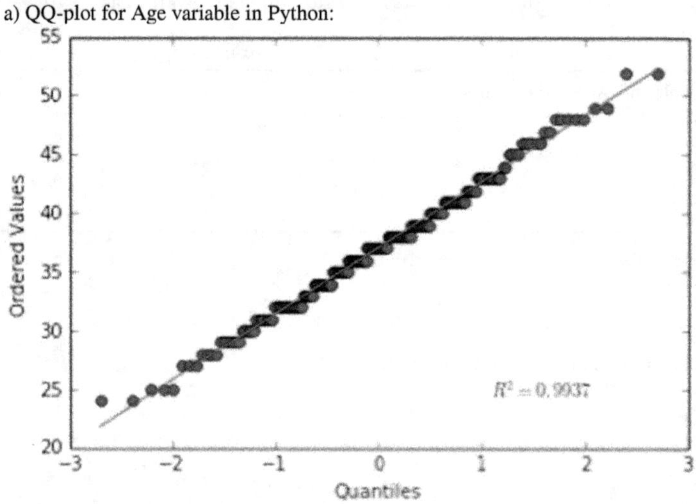

b) QQ-plot for Publications variable in Python:

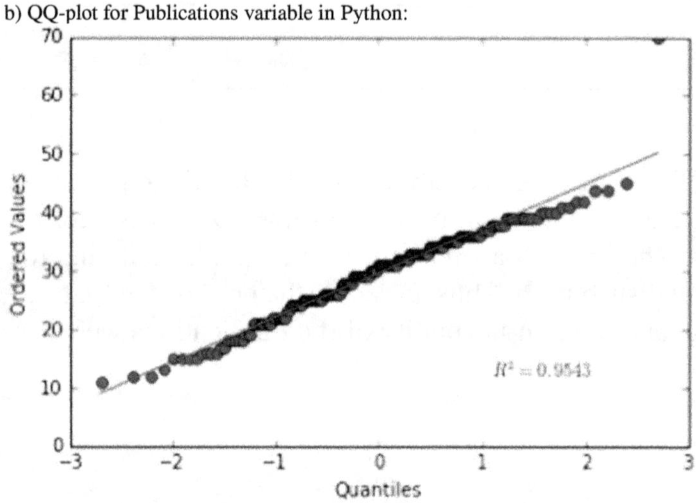

</td></tr>
</table>

Parametric Tests

As analyzed above, the null hypothesis was not rejected for Age variable. Thus, to understand the variation of this variable depending on other variables, a parametric test could be used.

Student's t-Test

The Student's t-test is used when the objective is to analyze the distribution of a numerical variable by a categorical variable with two-factor levels. For example, the variation of the Age with the Gender (male vs. female) is one of these cases.

When the null hypothesis states that there is no difference between the two population means (i.e., the difference equal to zero), the null and alternative hypothesis are often stated in the following form. So, the hypotheses are:

H_0: $\mu_1 = \mu_2$.
H_1: $\mu_1 \neq \mu_2$.

Here is an application of the Student's t-test in R and Python in Table 8 and Table 9.

The outputs shown in Tables 8 and 9 give the value of the test (t), the number of degrees of freedom (in R) and the *p*-value. Additionally, in R,

Table 8. R language: Student's t-test

	In R

Code	`### Student's t-test`
	`# Student's t-test with replacing the missing values by the mean of the variable`
	`t.test(na.aggregate(data.df[data.df$Gender=="male","Age"]), y = na.aggregate(data.df[data.df$Gender=="female","Age"]), var.equal=TRUE)`

Output	`Two Sample t-test`
	`data: na.aggregate(data.df[data.df$Gender == "male", "Age"]) and na.aggregate(data.df[data.df$Gender == "female", "Age"])`
	`t = -0.2855, df = 198, p-value = 0.7755`
	`alternative hypothesis: true difference in means is not equal to 0`
	`95 percent confidence interval:`
	`-1.810224 1.352318`
	`sample estimates:`
	`mean of x mean of y`
	`36.95495 37.18391`

Table 9. Python language: Student's t-test

	In Python

Code	`### Student's t-test`
	`# Student's t-test with replacing the missing values by the mean of the variable`
	`datadf['Age']=datadf['Age'].replace('nan',datadf['Age'].mean())`
	`print(stats.ttest_ind(datadf['Age'][datadf['Gender'] == "male"],datadf['Age'][datadf['Gender'] == "female"]))`

Output	`Ttest_indResult(statistic=-0.28330852883199081, pvalue=0.77723633340323406)`

some useful values are presented, namely the limits of 95% confidence interval. The *p*-value of the output is 0.78 > 0.05 (please be aware of the existence of small differences in the values obtained with R and with Python, probably due to intermediate rounding; however, they do not affect the final conclusions). Thus, in this case, there are no statistical differences, and the null hypothesis is not rejected. This means that the assumption of the equality of means of the two-factor levels is not rejected. It is possible to assume that male and female researcher's average age are approximately the same.

ANOVA

The one-way analysis of variance (ANOVA) is used to determine whether there are any significant differences between the means of three or more independent (unrelated) groups.

To apply an ANOVA test, there are three needed primary assumptions:

- The dependent variable is normally distributed in each group that is being compared in the one-way ANOVA.
- There is variances homogeneity. The population variances in each group are equal. Levene's Test is used for testing the Homogeneity of Variances.
- Independence of observations. This is mostly a study design issue and, as such, it will be needed to determine if they are independent.

The one-way ANOVA compares the means between the groups that the analyst is interested in and determines whether any of those means are significantly different from each other.

Specifically, the hypotheses are:

H$_0$: $\mu_1 = \mu_2 = \mu_3 = \ldots = \mu_k$.
H$_1$: The means are not all equal.

where μ = group mean and k is equal to the number of groups.

If, however, the one-way ANOVA returns a significant result, we accept the alternative hypothesis (H$_1$), which is that there are at least 2 group means that are significantly different from each other.

At this point, it is important to note that the one-way ANOVA is an omnibus statistic test and cannot tell which particular groups were significantly different from each other. It informs only that at least two groups are different. To determine which specific groups differ from each other, the reader needs to use a post hoc test.

In this context, to analyze the distribution of Age variable by different Tasks, the ANOVA test will be the most indicated. The normality tests show that Age variable presents a normal distribution. Therefore, the next step will be to analyze the Levene's test, to verify the homogeneity of variances.

Levene's test is used to test if k samples have equal variances. Equal variances across samples are called homogeneity of variance. Some statistical tests, for example, the analysis of variance, assume that variances are equal across groups or samples. The Levene's test can be used to verify that assumption.

The Levene test hypotheses are:

H$_0$: $\sigma_1^2 = \sigma_2^2 = \sigma_3^2 = \ldots = \sigma_k^2$.
H$_1$: $\sigma_i^2 \neq \sigma_j^2$ for at least one pair (i, j).

This test is presented in Table 10 and Table 11.

Table 10. R language: Levene's test

	In R

Code	`### Levene's test`
	`# Reading the package car, needed to apply Levene's test`
	`library(car)`
	`# Levene's test of the Age depending on the Tasks. Test with NA's substituted by the mean`
	`leveneTest(na.aggregate(data.df$Age) ~ Tasks, data.df)`

Output	`Levene's Test for Homogeneity of Variance (center = median)`
	`Df F value Pr(>F)`
	`group 2 4.7188 0.009959 **`
	`197`
	`---`
	`Signif. codes: 0 '***' 0.001 '**' 0.01 '*' 0.05 '.' 0.1 ' ' 1`

Table 11. Python language: Levene's test

	In Python

Code	### Levene's test
	# Levene's test of the Age depending of the Tasks. Test with NA's substituted by the mean
	print(stats.levene(datadf['Age'][datadf['Tasks']=="PhD_Student"],datadf['Age'][datadf['Tasks']=="Phd_Supervisor"],datadf['Age'][datadf['Tasks']=="Postdoctoral_research"], center = 'median'))

Output	LeveneResult(statistic=4.7188394172558903, pvalue=0.0099588685911839413)

With the Levene's test output, it is possible to verify, for a significance level of 5%, the null hypothesis is rejected ($p<0.0001<0.05$). Thus, no homogeneity of the variance should be assumed.

In the case of homogeneity of variances, the ANOVA test could be used without problems. However, if data has heterogeneity, one approach is to use the Welch correction as well as alternative post-hoc tests (i.e., a Games-Howell test instead of a Tukey post-hoc test). As, to our knowledge, ANOVA test for Welch's correction does not exist in Python. In this book, it was done using the R software.

The ANOVA test with Welch correction is presented in Table 12 and Table 13.

The ANOVA test gives a $p < 0.0001 < 0.05$. This allows concluding the rejection of the null hypothesis. Thus, it is possible to say that there are statistical differences between groups. However, as there are three-factor levels, post-hoc tests should be used. In a case of the variances homogeneity being violated, the Games-Howell test is more appropriate then Tukey test.

The Games-Howell test is used when variances are unequal and also takes into account unequal group sizes. Severely unequal variances can lead to increased Type I error, and, with smaller sample sizes, the more moderate

Table 12. R language: ANOVA test with Welch's correction

	In R
Code	### ANOVA test with Welch's correction oneway.test(na.aggregate(data.df$Age)~Tasks, data = data.df, na.action=na.omit, var.equal=FALSE)
Output	One-way analysis of means (not assuming equal variances) data: na.aggregate(data.df$Age) and Tasks F = 173.5291, num df = 2.000, denom df = 121.464, p-value < 2.2e-16

Table 13. Python language: ANOVA test with Welch's correction

	In Python

Code	### One-Way ANOVA with R for Welch's correction
	# Import packages
	from pandas import *
	import rpy2 as rpy2
	from rpy2.robjects.packages import importr
	import rpy2.robjects as ro
	import pandas.rpy.common as com
	# Changing R's directory to where is the data
	ro.r('setwd("C:/Users/Rui Sarmento/Documents/Livro Cybertech/Dados e Code")')
	# Reading the data with R
	ro.r('data_df <- read.csv("data.csv",sep=";")')
	# Get NA values equal to Age's mean with function na.aggregate and *zoo* package
	ro.r('library(zoo)')
	ro.r('library(stats)')
	# ANOVA with Welch correction (var.equal = FALSE)
	print(ro.r('oneway.test(na.aggregate(data_df$Age)~Tasks, data = data_df, na.action=na.omit, var.equal=FALSE)'))

Output	One-way analysis of means (not assuming equal variances) data: na.aggregate(data.df$Age) and Tasks F = 173.5291, num df = 2.000, denom df = 121.464, p-value < 2.2e-16

differences in-group variance can result in increases in Type I error. The Games-Howell test, which is designed for unequal variances, is based on Welch's correction. See Table 14 and Table 15.

The given output is very complete. The descriptive analysis of the Age variable by each task is represented. Also, the results of the Levene's test, the ANOVA test, and the post-hoc tests are present in this output. While ANOVA can tell the reader whether groups in the sample differ, it cannot tell the reader which groups differ from each other. That is, if the results of ANOVA indicate that there is a significant difference among the groups, the obvious question becomes: Which groups in this sample differ significantly? It is not likely that all groups differ when compared to each other, only that a handful has significant differences. Post-hoc tests can clarify the reader of which groups among the sample have significant differences. In this sense, the Student's t-test compares the age of Ph.D. Students vs. Ph.D. Supervisor, Ph.D. Students vs. Postdoctoral researchers and Ph.D. Supervisor vs. Post-doctoral researchers. In all cases, $p<0.001<0.05$, showing significant differences between all different tasks.

Looking again at descriptive results at the beginning of this output, it is possible to see that, on average, Ph.D. Students have 32.05 years old,

Table 14. R language: Games-Howell test for multiple comparisons

	In R

Code	### Games-Howell test for multiple comparisons
	# Substitution of the missing values by the mean of the variable
	data.df.new$Age <- na.aggregate(as.numeric(data.df$Age), by="Age", FUN = mean)
	# Reading the package userfriendlyscience, needed to apply Games-Howell test
	library(userfriendlyscience)
	# Games-Howell test
	oneway(y=data.df$Age, x = data.df$Tasks, posthoc="games-howell", means=T, fullDescribe=T, levene=T,
	plot=T, digits=2, pvalueDigits=3, conf.level=.95)

Output	### Means for y (Age) separate for each level of x (Tasks): x: PhD_Student n mean sd median trimmed mad min max range skew kurtosis se 78 32.05 3.55 32 32.14 2.97 24 39 15 -0.19 -0.41 0.4 -- x: Phd_Supervisor n mean sd median trimmed mad min max range skew kurtosis se 56 43.48 3.47 43 43.35 2.97 36 52 16 0.43 -0.36 0.46 -- x: Postdoctoral_research n mean sd median trimmed mad min max range skew kurtosis se 66 37.52 2.32 37.06 37.41 2.88 33 43 10 0.41 -0.21 0.28 ### Oneway Anova for y=Age and x=Tasks (groups: PhD_Student, Phd_Supervisor, Postdoctoral_research) Eta Squared: 95% CI = [0.62; 0.73], point estimate = 0.68 SS Df MS F p Between groups (error + effect) 4280.24 2 2140.12 212.91 <.001 Within groups (error only) 1980.15 197 10.05 ### Levene's test: Levene's Test for Homogeneity of Variance (center = mean) Df F value Pr(>F) group 2 5.4902 0.004783 ** 197 --- Signif. codes: 0 '***' 0.001 '**' 0.01 '*' 0.05 '.' 0.1 ' ' 1 ### Post hoc test: games-howell t df p PhD_Student:Phd_Supervisor 18.63 120.22 <.001 PhD_Student:Postdoctoral_research 11.09 133.83 <.001 Phd_Supervisor:Postdoctoral_research 10.96 93.16 <.001

Postdoctoral Researchers have 37.52 years old, and Ph.D. Supervisors have 43.48 years old.

Although Games-Howell is more appropriate when there are differences between groups without homogeneity of variances, the Tukey test is also present in this book. The objective is to show to the reader how he/she could use the test. Note that Tukey test should only be utilized in case of homo-

Table 15. Python language: Games-Howell test for multiple comparisons

	In Python

Code	### Games-Howell test for multiple comparisons (asking R)
	# See https://sites.google.com/site/aslugsguidetopython/data-analysis/pandas/calling-r-from-python
	# Import packages
	from pandas import *
	import rpy2 as rpy2
	from rpy2.robjects.packages import importr
	import rpy2.robjects as ro
	import pandas.rpy.common as com
	# Changing R's directory to where is the data
	ro.r('setwd("C:/Users/Rui Sarmento/Documents/Livro Cybertech/Dados e Code")')
	# Reading the data with R
	ro.r('data_df <- read.csv("data.csv",sep=";")')
	# Reading the R's package zoo, needed to apply na.aggregate
	ro.r('library(zoo)')
	# Store the data in an auxiliary variable for editing
	ro.r('data_df_new <- data_df')
	# Substitution of the missing values by the mean of the variable
	ro.r('data_df_new$Age <- na.aggregate(as.numeric(data_df$Age), by="Age", FUN = mean)')
	# Reading the R's package userfriendlyscience, needed to apply Games-Howell test
	# this time the loading is done with the importr function
	science = importr('userfriendlyscience')
	# Games-Howell test
	print(ro.r('oneway(y=data_df_new$Age, x = data_df$Tasks, posthoc="games-howell", means=T, fullDescribe=T, levene=T,plot=T, digits=2, pvalueDigits=3, conf.level=0.95)'))

Output	### Means for y (Age) separate for each level of x (Tasks): x: PhD_Student n mean sd median trimmed mad min max range skew kurtosis se 78 32.05 3.55 32 32.14 2.97 24 39 15 -0.19 -0.41 0.4 -- x: Phd_Supervisor n mean sd median trimmed mad min max range skew kurtosis se 56 43.48 3.47 43 43.35 2.97 36 52 16 0.43 -0.36 0.46 -- x: Postdoctoral_research n mean sd median trimmed mad min max range skew kurtosis se 66 37.52 2.32 37.06 37.41 2.88 33 43 10 0.41 -0.21 0.28 ### Oneway Anova for y=Age and x=Tasks (groups: PhD_Student, Phd_Supervisor, Postdoctoral_research) Eta Squared: 95% CI = [0.62; 0.73], point estimate = 0.68 SS Df MS F p Between groups (error + effect) 4280.24 2 2140.12 212.91 <.001 Within groups (error only) 1980.15 197 10.05 ### Levene's test: Levene's Test for Homogeneity of Variance (center = mean) Df F value Pr(>F) group 2 5.4902 0.004783 ** 197 --- Signif. codes: 0 '***' 0.001 '**' 0.01 '*' 0.05 '.' 0.1 ' ' 1 ### Post hoc test: games-howell t df p PhD_Student:Phd_Supervisor 18.63 120.22 <.001 PhD_Student:Postdoctoral_research 11.09 133.83 <.001 Phd_Supervisor:Postdoctoral_research 10.96 93.16 <.001

geneity of variances. Tukey's HSD test is a post-hoc test, meaning that it is performed after an analysis of variance (ANOVA) test. This means that to maintain integrity, a statistician should not perform Tukey's HSD test unless he/she has first performed an ANOVA analysis. In statistics, post-hoc tests are used only for further data analysis. These types of tests are not pre-planned. In other words, the reader should have no plans to use Tukey's HSD test before he/she collects and analyzes the data first. See Table 16 and Table 17.

The last column of the Tukey test output gives the p-value (in R) or the rejection (equal to true or false in Python). In all comparisons, the p-value

Table 16. R language: Tukey test

	In R

Code	`### Tukey test`
	`data.anova <- aov(na.aggregate(data.df$Age)~Tasks, data = data.df)`
	`TukeyHSD(data.anova)`

Output	`Tukey multiple comparisons of means`
	`95% family-wise confidence level`
	`Fit: aov(formula = na.aggregate(data.df$Age) ~ Tasks, data = data.df)`
	`$Tasks`
	`diff lwr upr p adj`
	`Phd_Supervisor-PhD_Student 11.430861 10.119485 12.742237 0`
	`Postdoctoral_research-PhD_Student 5.465553 4.213340 6.717766 0`
	`Postdoctoral_research-Phd_Supervisor -5.965308 -7.325594 -4.605022 0`

Table 17. Python language: Tukey test

	In Python

Code	`### Tukey test`
	`# Import packages`
	`from statsmodels.stats.multicomp import pairwise_tukeyhsd`
	`# Exchanging "nan" values by the mean of the variable`
	`datadf['Age']=datadf['Age'].replace('nan',datadf['Age'].mean())`
	`# Tukey test`
	`tukey = pairwise_tukeyhsd(endog=datadf['Age'],`
	`groups=datadf['Tasks'], alpha=0.05)`
	`print(tukey.summary ())`

Output	`Multiple Comparison of Means - Tukey HSD,FWER=0.05`
	`==================================================================`
	`group1 group2 meandiff lower upper reject`
	`------------------------------------------------------------------`
	`PhD_Student Phd_Supervisor 11.4309 10.1194 12.7423 True`
	`PhD_Student Postdoctoral_research 5.4656 4.2133 6.7179 True`
	`Phd_Supervisor Postdoctoral_research -5.9653 -7.3257 -4.6049 True`
	`------------------------------------------------------------------`

is near zero. This means that the hypothesis of the same mean of ages between groups is rejected and, therefore, there are significant differences between them.

Non-Parametric Tests

Nonparametric tests are useful to test whether group means or medians are distributed the same way across groups. In these types of tests, we rank (or place in order) each observation from our data set. Nonparametric tests are widely used when the reader does not know whether data follows a normal distribution, or have confirmed the data does not follow a normal distribution. On the other side, hypothesis tests are parametric tests based on the assumption that the population follows a normal distribution with a set of parameters.

In general, conclusions drawn from non-parametric methods are not as robust as the parametric ones. However, as non-parametric methods make fewer assumptions, they are more flexible, more robust, and applicable to non-quantitative data.

As analyzed at the beginning of this chapter, the hypothesis of normality was rejected for the Publications variable. Thus, some non-parametric tests are used to analyze this variable.

Mann-Whitney Test

The Mann-Whitney U test is the alternative test to the independent sample Student's t-test. It is a nonparametric test that allows two groups or conditions or treatments to be compared without making the assumption that values are normally distributed. Thus, for example, we might compare the speed at which two different groups of people can run 100 meters, where one group has trained for six weeks and the other has not.

Since Mann-Whitney U test is a non-parametric test, it does not assume any assumptions related to the distribution. There are, however, some assumptions that are considered:

- The sample drawn from the population is random.
- Independence within the samples and mutual independence is considered.
- The ordinal measurement scale is assumed.

The hypotheses by this test are:

H$_0$: The medians of the two samples are identical.
H$_1$: The medians of the two samples are different.

So, to analyze the distribution of the Publications variable, depending on the gender of the researchers, Mann-Whitney test should be used. It is presented in Table 18 and Table 19.

Mann-Whitney test output shows the value of the test (W in R or statistic in Python) and the *p*-value. In this case, *p*-value=0.2887 > 0.05, which leads to not reject the null hypothesis. Thus, it is possible to say that the hypothesis of the medians of the two samples being identical is not rejected and, therefore, male and female have a similar number of publications.

Kruskal-Wallis Test

Kruskal-Wallis test is a nonparametric test and is used when the assumptions of ANOVA are not met. They both assess for significant differences on a continuous dependent variable by a grouping independent variable (with three or more groups). In ANOVA test, we assume that distribution of each group is normally distributed and there is approximately equal variance on the scores for each group. However, in the Kruskal-Wallis test, we do not have

Table 18. R language: Mann-Whitney test

	In R
Code	`### Mann-Whitney test` `wilcox.test(na.aggregate(data.df[data.df$Gender == "male", "Publications"]), y = na.aggregate(data.df[data.df$Gender == "female", "Publications"]))`

Output	`Wilcoxon rank sum test with continuity correction`
	`data: na.aggregate(data.df[data.df$Gender == "male", "Publications"]) and na.aggregate(data.df[data.df$Gender == "female", "Publications"])`
	`W = 4485, p-value = 0.2887`
	`alternative hypothesis: true location shift is not equal to 0`

Table 19. Python language: Mann-Whitney test

	In Python
Code	`### Mann-Whitney test` `print(stats.mannwhitneyu(datadf['Publications'][datadf['Gender']== "male"],datadf['Publications'][datadf['Gender']=="female"]))`
Output	`MannwhitneyuResult(statistic=4485.0, pvalue=0.28871061294321942)`

any of these assumptions. Like all non-parametric tests, the Kruskal-Wallis test is not as powerful as ANOVA.

The hypotheses are:

H_0: The samples are from identical populations.
H_1: The samples come from different populations.

Thus, to analyze the distribution of the Publications variable depending on the tasks of the researchers, Kruskal-Wallis test should be used. It is presented in Table 20 and Table 21.

Kruskal-Wallis test output gives the value of the test (Kruskal-Wallis chi-squared in R or statistic in Python), the number of degrees of freedom (only in R) and the *p*-value. In the case of the number of publications depending on the tasks, p=7.138e-09 < 0.05 and, therefore, the null hypothesis is rejected. Thus, it is possible to say that different tasks have a different number of publications.

Table 20. R language: Kruskal-Wallis test

	In R

Code	`### Kruskal-Wallis test`
	`kruskal.test(list(data.df[data.df$Tasks=="PhD_`
	`Student","Publications"],data.df[data.df$Tasks=="Postdoctoral_`
	`research","Publications"],data.df[data.df$Tasks=="Phd_`
	`Supervisor","Publications"]))`

Output	`Kruskal-Wallis rank sum test`
	`data: list(data.df[data.df$Tasks == "PhD_Student", "Publications"],`
	`data.df[data.df$Tasks == "Postdoctoral_research", "Publications"],`
	`data.df[data.df$Tasks == "Phd_Supervisor", "Publications"])`
	`Kruskal-Wallis chi-squared = 37.5156, df = 2, p-value = 7.138e-09`

Table 21. Python language: Kruskal-Wallis test

	In Python

Code	`### Kruskal-Wallis test`
	`from scipy.stats.mstats import kruskalwallis`
	`print(kruskalwallis(datadf['Publications'][datadf['Tasks']=="PhD_`
	`Student"],datadf['Publications'][datadf['Tasks']=="Phd_Superv`
	`isor"],datadf['Publications'][datadf['Tasks']=="Postdoctoral_`
	`research"]))`

Output	`KruskalResult(statistic=37.515595218096237,` `pvalue=7.1382541375268384e-09)`

However, which groups have a big difference in the number of published papers is unknown. To know the group responsible for the significant differences, it is needed to make multiple comparisons. In ANOVA test, these comparisons are made with the post-hoc test. In Kruskal-Wallis test, there are no post-hoc tests. So, the comparisons should be made with Mann-Whitney test, in Table 22 and Table 23.

In the outputs shown in Tables 22 and 23, the results of the three Mann-Whiney tests are presented for multiple comparisons. The resulting *p*-values are 0.00109, 2.202e-09, and 0.002125. All of these values are lower than 0.05 and, therefore, the three null hypotheses are rejected. The obtained results allow concluding that there are significant differences between all tasks. If the reader needs to know which tasks have more publications, a descriptive analysis of them should be explored.

Crosstabs and Chi-Square Test

To summarize a single categorical variable, we use frequency tables. To summarize the relationship between two categorical variables, we use a

Table 22. R language: multiple comparisons with Mann-Whitney test

<table>
<tr><th></th><th>In R</th></tr>
<tr><td>Code</td><td>

```
### Multiple comparisons with Mann-Whitney test
# Mann-Whitney test for Ph.D. Students vs. Postdoctoral Researchers
wilcox.test(data.df[data.df$Tasks == "PhD_Student", "Publications"], y
= data.df[data.df$Tasks == "Postdoctoral_research", "Publications"])
# Mann-Whitney test for Ph.D. Students vs. Ph.D. Supervisors
wilcox.test(data.df[data.df$Tasks == "PhD_Student", "Publications"], y
= data.df[data.df$Tasks == "Phd_Supervisor", "Publications"])
# Mann-Whitney test for Postdoctoral Researchers vs. Ph.D. Supervisors
wilcox.test(data.df[data.df$Tasks == "Postdoctoral_research",
"Publications"], y = data.df[data.df$Tasks == "Phd_Supervisor",
"Publications"])
```

</td></tr>
<tr><td>Output</td><td>

```
Wilcoxon rank sum test with continuity correction
data: data.df[data.df$Tasks == "PhD_Student", "Publications"] and
data.df[data.df$Tasks == "Postdoctoral_research", "Publications"]
W = 1760, p-value = 0.00109
alternative hypothesis: true location shift is not equal to 0
Wilcoxon rank sum test with continuity correction
data: data.df[data.df$Tasks == "PhD_Student", "Publications"] and
data.df[data.df$Tasks == "Phd_Supervisor", "Publications"]
W = 859, p-value = 2.202e-09
alternative hypothesis: true location shift is not equal to 0
Wilcoxon rank sum test with continuity correction
data: data.df[data.df$Tasks == "Postdoctoral_research",
"Publications"] and data.df[data.df$Tasks == "Phd_Supervisor",
"Publications"]
W = 1250.5, p-value = 0.002125
alternative hypothesis: true location shift is not equal to 0
```

</td></tr>
</table>

Table 23. Python language: multiple comparisons with Mann-Whitney test

	In Python

Code	`### Multiple comparisons with Mann-Whitney test`
	`# Mann-Whitney test for Ph.D. Students vs. Postdoctoral Researchers`
	`print(stats.mannwhitneyu(datadf['Publications'][datadf['Tasks']=="PhD_Student"],datadf['Publications'][datadf['Tasks']=="Postdoctoral_research"]))`
	`# Mann-Whitney test for Ph.D. Students vs. Ph.D. Supervisors`
	`print(stats.mannwhitneyu(datadf['Publications'][datadf['Tasks']=="PhD_Student"],datadf['Publications'][datadf['Tasks']=="Phd_Supervisor"]))`
	`# Mann-Whitney test for Postdoctoral Researchers vs. Ph.D. Supervisors`
	`print(stats.mannwhitneyu(datadf['Publications'][datadf['Tasks']=="Phd_Supervisor"],datadf['Publications'][datadf['Tasks']=="Postdoctoral_research"]))`

Output	`MannwhitneyuResult(statistic=1760.0, pvalue=0.001090154886338399)`
	`MannwhitneyuResult(statistic=859.0, pvalue=2.2020601202061945e-09)`
	`MannwhitneyuResult(statistic=2445.5, pvalue=0.0021248086990779723)`

cross-tabulation (also called the contingency table). A cross-tabulation (or crosstab) is a table that depicts the number of times each of the possible category combinations occurred in the sample data.

The chi-square test for independence also called Pearson's chi-square test, or the chi-square test of association is used to test independence between the row and column variables. Independence means that knowing the value of the row variable does not change the probabilities of the column variable (and vice versa). Another way of looking at independence is to say that the row percentages (or column percentages) remain constant from row to row (or column to column).

When a chi-square test for independence is chosen to analyze our data, it is needed to make sure that the data we want to analyze "passes" two assumptions. It is necessary to do this because it is only appropriate to use a chi-square test for independence if the data gives these two assumptions. If it does not, a chi-square test for independence cannot be used. These two assumptions are:

- The two variables should be measured at an ordinal or nominal level (i.e., categorical data).
- The two variables should consist of two or more categorical, independent groups.

Also, the hypotheses of this test are:

H_0: The variables are independent.

H_1: The variables are not independent.

Some comparisons between categorical variables are presented in Table 24 and Table 25.

To compare two or more categorical variables in R, it is first needed to do a crosstab and, after that, to apply the chi-squared test to the crosstab. Thus, comparing Gender with R_user, the reader verifies, in the output, that

Table 24. R language: Chi-squared test for Gender vs. R_user

	In R

Code	`### Chi-squared test for Gender vs. R_user`
	`# Contingency table of Gender vs. R_user`
	`tbl <- table(data.df$Gender, data.df$R_user)`
	`# Chi-squared test`
	`chisq.test(tbl)`
	`# Asking expected values`
	`chisq.test(tbl)$expected`

Output	`# Contingency table`
	`no yes`
	`female 54 33`
	`male 37 76`
	`# Chi-squared test`
	`Pearson's Chi-squared test with Yates' continuity correction`
	`data: tbl`
	`X-squared = 15.8851, df = 1, p-value = 6.731e-05`
	`# Expected values`
	`no yes`
	`female 39.585 47.415`
	`male 51.415 61.585`

Table 25. Python language: Chi-squared test for Gender vs. R_user

	In Python

Code	`### Chi-squared test for Gender vs. R_user`
	`# Import packages`
	`import scipy.stats as stats`
	`import numpy as np`
	`import pandas as pd`
	`# Chi-squared test`
	`table_r = datadf.pivot_table(index='Gender',columns='R_user',`
	`values = 'id',aggfunc='count')`
	`print(stats.chi2_contingency(table_r))`

Output	`print(stats.chi2_contingency(table_r))`
	`(15.885131778780307, 6.7305396658499117e-05, 1, array([[ 39.585,`
	`47.415],`
	`[ 51.415, 61.585]]))`

there are 54 female non-users and 33 female users. On the opposite, the number of male non-users is 37, and male users count 76.

When the chi-squared test is applied, the test value, degrees of freedom and *p*-value are given in the output. In this case, p=6.731e-05 < 0.05. Thus, the null hypothesis is rejected, i.e., the hypothesis of independence of the variables is rejected. In this context, it is possible to conclude that Gender and R_user are dependents. By the previous results of the crosstab and the *p*-value is clear that male is more inclined to use R than female.

The same comparison for Gender vs. Python_user is also available. See Table 26 and Table 27.

Similar to the comparison of Gender vs. R_user, in Gender vs. Python_user, the *p*-value is lower than 0.05. Here, the null hypothesis is also rejected. In this case, it is possible to conclude that female is more inclined to use Python than male.

Regarding the last comparison for categorical variables, i.e., Gender vs. Tasks, we have the analyses in Table 28 and Table 29.

The previous output shows the Gender vs. Tasks crosstab and corresponding chi-square test. In this case, there are 34 female Ph.D. Students, 24 female Ph.D. Supervisor, and 29 female Postdoctoral researchers. Additionally, there are 44 male Ph.D. Students, 32 male PhD Supervisor, and 37 male Postdoctoral researchers. The *p*-value resulting from the chi-squared test is p=0.9926.

Table 26. R language: Chi-squared test for Gender vs. Python_user

	In R

Code	`### Chi-squared test for Gender vs. Python_user`
	`# Contingency table of Gender vs. Python_user`
	`tbl <- table(data.df$Gender, data.df$Python_user)`
	`# Chi-squared test`
	`chisq.test(tbl)`
	`# Asking expected values`
	`chisq.test(tbl)$expected`

Output	`# Contingency table`
	`no yes`
	`female 26 60`
	`male 66 47`
	`# Chi-squared test`
	`Pearson's Chi-squared test with Yates' continuity correction`
	`data: tbl`
	`X-squared = 14.4817, df = 1, p-value = 0.0001415`
	`# Expected values`
	`no yes`
	`female 39.75879 46.24121`
	`male 52.24121 60.75879`

Table 27. Python language: Chi-squared test for Gender vs. Python_user

	In Python

Code	`### Chi-squared test for Gender vs. Python_user`
	`# Import packages`
	`import scipy.stats as stats`
	`import numpy as np`
	`import pandas as pd`
	`# Chi-squared test`
	`table_py = datadf.pivot_table(index='Gender',columns='Python_user', values = 'id',aggfunc='count')`
	`print(stats.chi2_contingency(table_py))`

Output	`print(stats.chi2_contingency(table_py))`
	`(14.481674230676045, 0.00014152973642310866, 1, array([[ 39.75879397, 46.24120603],`
	`[ 52.24120603, 60.75879397]]))`

Table 28. R language: Chi-squared test for Gender vs. Tasks

	In R

Code	`### Chi-squared test for Gender vs. Tasks`
	`# Contingency table of Gender vs. Tasks`
	`tbl <- table(data.df$Gender, data.df$Tasks)`
	`# Chi-squared test`
	`chisq.test(tbl)`
	`# Asking expected values`
	`chisq.test(tbl)$expected`

Output	`# Contingency table`
	`PhD_Student Phd_Supervisor Postdoctoral_research`
	`female 34 24 29`
	`male 44 32 37`
	`# Chi-squared test`
	`Pearson's Chi-squared test`
	`data: tbl`
	`X-squared = 0.0149, df = 2, p-value = 0.9926`
	`# Expected values`
	`PhD_Student Phd_Supervisor Postdoctoral_research`
	`female 33.93 24.36 28.71`
	`male 44.07 31.64 37.29`

Hence, the null hypothesis is not rejected, and it is possible to claim that Gender and Tasks variables are independent.

Correlations

Correlation is a bivariate analysis that measures the strengths of association between two variables. In statistics, the correlation coefficient varies between +1 and -1. When the correlation coefficient lies around $\pm$ 1, it is said to be

Table 29. Python language: Chi-squared test for Gender vs. Tasks

	In Python

Code	`### Chi-squared test for Gender vs. Tasks`
	`# Import packages`
	`import scipy.stats as stats`
	`import numpy as np`
	`import pandas as pd`
	`# Chi-squared test`
	`table_tasks = datadf.pivot_table(index='Gender',columns='Tasks', values = 'id',aggfunc='count')`
	`print(stats.chi2_contingency(table_tasks))`

Output	`print(stats.chi2_contingency(table_tasks))`
	`(0.014856468930316906, 0.99259928668180353, 2, array([[ 33.93, 24.36, 28.71],`
	`[ 44.07, 31.64, 37.29]]))`

a perfect degree of association between the two variables. As the correlation coefficient value goes towards 0, the relationship between the two variables will be weaker. Usually, in statistics, three types of correlations are used: Pearson correlation, Kendall rank correlation, and Spearman correlation.

- **Pearson R Correlation:** Pearson r correlation is widely employed in statistics to measure the degree of the relationship between linear related variables. For Pearson r correlation, both variables should normally be distributed. Other assumptions include linearity and homogeneity.
- **Kendall Rank Correlation:** Kendall rank correlation is a non-parametric test that measures the strength of dependence between two variables.
- **Spearman Rank Correlation:** Spearman rank correlation is a non-parametric test that is used to gauge the degree of association between two variables. It was developed by Spearman. Thus, it is called the Spearman rank correlation. Spearman rank correlation test does not assume any assumptions about the distribution of the data and is the proper correlation analysis when the variables are measured on a scale that is at least ordinal. The assumptions of Spearman rho correlation are that data must be at least ordinal, and scores on one variable must be monotonically related to the other variable.

To analyze the two numerical variables in the study, correlations analysis should be applied. Since Publications variable does not have a normal distribution, Spearman's correlation is more appropriate. See Table 30 and Table 31.

Table 30. R language: Spearman's Correlations

	In R

Code	`### Spearman's Correlations`
	`cor.test(data.df$Publications, na.aggregate(data.df$Age),`
	`method="spearman")`

Output	`Spearman's rank correlation rho`
	`data: data.df$Publications and na.aggregate(data.df$Age)`
	`S = 597192.2, p-value < 2.2e-16`
	`alternative hypothesis: true rho is not equal to 0`
	`sample estimates:`
	`rho`
	`0.5520947`

Table 31. Python language: Spearman's Correlations

	In Python
Code	`### Spearman's Correlations` `print(stats.spearmanr(datadf['Publications'],datadf['Age']))`
Output	`SpearmanrResult(correlation=0.55209466617812353,` `pvalue=2.3689996688336485e-17)`

The outputs in Tables 30 and 31 show the value of the correlation (rho in R or correlation in Python) and the *p*-value. The value of the correlation suggests a moderate positive relationship (rho=0.55). The laid hypothesis of independence of variables (rho equal to zero) is rejected. Thus, the Age and Publications variables are positively correlated, i.e., senior researchers have more publications, or vice-versa.

CONCLUSION

To understand the simple difference between descriptive and inferential statistics, all the reader needs to remember is that descriptive statistics summarize the current dataset, and inferential statistics aim to draw conclusions about an additional population outside of the dataset.

The central concepts explained in this chapter are:

- Statistical Tests:
    - Shapiro-Wilk test,
    - Kolmogorov-Smirnov test,
    - Student's t-test,

Table 32. Synthesizing which statistical test should be used

<table>
<tr><th colspan="2"></th><th colspan="3">Parametric</th><th>Non-Parametric</th></tr>
<tr><td colspan="2">Correlation</td><td colspan="3">Pearson</td><td>Spearman</td></tr>
<tr><td rowspan="4">Statistical test</td><td>2 groups</td><td colspan="3">Student's t-test</td><td>Mann-Whitney test</td></tr>
<tr><td rowspan="3">> 2 groups</td><td>Assumptions</td><td>Statistical Test</td><td>Post-Hoc Tests</td><td rowspan="3">Kruskal-Wallis test</td></tr>
<tr><td>Homogeneity of Variance (Levene's test)</td><td>One-way ANOVA</td><td>Tukey test</td></tr>
<tr><td>No Homogeneity of Variance (Levene's test)</td><td>ANOVA with Welch's correction</td><td>Games-Howell test</td></tr>
</table>

  - ANOVA,
  - Levene's test,
  - Welch correction,
  - Games-Howell test,
  - Tukey test,
  - Mann-Whitney test,
  - Kruskal-Wallis test,
  - Chi-square test.
- Correlations:
  - Pearson,
  - Spearman,
  - Kendall.

To synthesize which statistical test should be used depending on the data type. Independent samples are shown in Table 32.

Chapter 6
Introduction to Linear Regression

INTRODUCTION

In statistical modeling, regression analysis is a statistical process for estimating the relationships among variables. It includes many techniques for modeling and analyzing several variables when the focus is on the relationship between a dependent variable and one or more independent variables i.e. the predictors. The terminology "dependent variable" might suggest a cause-effect relationship between variables but, in this case, the language is applied more significantly to the discovery of a mathematical model.

The "regression" term was first applied to this type of mathematical studies by Sir Francis Galton in 1885. Sir Galton study demonstrated that the height of the sons is not related to the father's height but has a tendency to "regress" or approach the average population height.

More specifically, regression analysis helps the reader understand how the dependent variable changes when any of the independent variables is varied. When two or more independent values are available, this is true when the other independent variables are fixed. Thus, regression analysis estimates the average value of the dependent variable when the independent variables are fixed. However, other approaches might be useful like, for example, the focus might be on other location than the average, a quantile or other location parameter of the conditional distribution of the dependent variable given the fixed independent variables.

DOI: 10.4018/978-1-68318-016-6.ch006

Therefore, the estimation target is a function of the independent variables called regression function. Regression analysis is widely used for prediction and forecasting. Regression analysis is also used to understand which among the independent variables are acquainted with the dependent variable. In limited circumstances, regression analysis can be used to infer causal relationships between the independent and dependent variables. Nonetheless, caution has to be taken since correlation might not signify causality. For example, consider that a Ph.D. Supervisor might be interested in measuring the influence of hours of study, annual income, number of brothers and sisters, marital status in the students' scores. Although some of these variables might appear to have some causality in the model, this is sometimes not clear and even unnecessary to build the regression model.

Regression analysis techniques are varied. Nonetheless, in this chapter, we will present only the basic analysis.

R VS. PYTHON

To make a linear regression model, the variability of Publications variable was studied when related with the Gender, R_user, Python_user, and Age variables.

Linear Regression Model

In linear regression model, the functional relationship between the dependent variable and the independent variables $\left(X_i; i = 1,\ldots,p\right)$ are (Maroco, 2011):

$$Y_j = \beta_0 + \beta_1 X_{1j} + \beta_2 X_{2j} + \ldots + \beta_p X_{pj} + \varepsilon_j, \left(j = 1,\ldots,n\right)$$

In this model, β_i are the regression coefficients and ε_j is the errors or residuals of the model. β_0 is the y-intercept, and β_i represents the partial slopes (i.e. a measure of the influence of X_i in Y, i.e., the Y variation per variation unit of X_i). The term ε_j (errors or residuals of the model) reflects the measurement errors and the natural variation in Y. If there is only one independent variable, the model is called simple linear regression model. If the model has more than one independent variable, it is called multiple linear regression model.

Considering that the population is not defined, the linear regression analysis should start with the estimation of the regression coefficients from a representative sample of the population under study, using the estimators

$$\hat{Y}_j = \hat{\beta}_0 + \hat{\beta}_1 X_{1j} + \hat{\beta}_2 X_{2j} + \ldots + \hat{\beta}_p X_{pj}$$

producing sample estimates $b_0, b_1, \ldots, b_p$ of the population's parameters $\beta_0, \beta_1, \ldots, \beta_p$. The commonly used methods for estimation of these coefficients are mostly very laborious. Most software has extensive modules for linear regression. Thus, they eliminate the task of estimating these parameters; its detailed presentation will not be available in this book.

Inference in Linear Regression

After $b_0, b_1, \ldots, b_p$ estimates being found, the reader should proceed to the evaluation of the quantitative influence of independent variables on the dependent variable in the sample.

ANOVA for Regression

The goal now is to evaluate, from sample estimates if, in fact, in the population, some of the independent variables may or may not influence the dependent variable, i.e., if the adjusted model is significant or not (Marôco, 2011). This hypothesis could be written as:

H$_0$: $\beta_1 = \beta_2 = \ldots = \beta_p = 0$.
H$_1$: $\exists i : \beta_i \neq 0 \left(i = 1, \ldots, p\right)$.

To test these hypotheses, the total variation in the response Y is divided into two components: one term is the variation in mean response, and the other term is the residual value. This gives the equation

$$\sum_{j=1}^{n}\left(Y_j - \bar{Y}\right)^2 = \sum_{j=1}^{n}\left(\hat{Y}_j - \bar{Y}\right)^2 + \sum_{j=1}^{n}\left(Y_j - \hat{Y}_j\right)^2$$

This equation may also be written as $SST = SSM + SSE$, where *SS* is the notation for the sum of squares and *T*, *M*, and *E* are the notation for total, model, and error, respectively.

For multiple linear regression, the statistic $MSM\ /\ MSE$ has an *F* distribution with $(DFM, DFE) = (p, n-p-1)$ degrees of freedom, where *p* is the number of independent variables in the model and *n* is the number of variables.

Hence, if *p*-value < α, H_0 is rejected and it is possible to conclude that, at least one independent variable, has a significant effect on the variance of the dependent variable. This does not mean that the independent variable is the cause of the dependent variable. It can only be said that the adjusted model to the data is significant. However, it should be checked if all or only some independent variables influence the variation of the dependent variable.

Tests on the Regression Coefficients

Similar to the variance analysis for two or more population's averages, by rejecting H_0 in the regression ANOVA, we can only conclude that at least one β_i is significantly different from zero. To determine which of the $\beta_i \left(i = 1, \dots, p\right)$ is nonzero is necessary to carry out multiple β_i tests. The statistics hypotheses are:

$H_{0:}$ $\beta_i = k$
H_1: $\beta_i \neq k\left(i = 1, \dots, p\right)$ and $k = 0$ in most software.

To test the presented hypotheses, the statistic test has Student's t-distribution with $\left(n - p - 1\right)$ degrees of freedom. If *p*-value < α, H_0 is rejected. Please note that Student's t-test is only valid for each variable, one at a time. The extrapolation of which variables simultaneously influence on the dependent variable is not valid.

Coefficient of Determination

The coefficient of determination, denoted by R^2 or r^2 is a number that indicates the proportion of variance in the dependent variable that is predictable from the independent variable.

It is a statistic used in the context of statistical models whose primary purpose is either the prediction of future outcomes or the testing of hypotheses, by other related information. It provides a measure of how well observed outcomes are replicated by the model, based on the proportion of total variation of the results explained by the model.

The square of the sample correlation is equal to the ratio of the model squares sum to the total sum of squares:

$$r^2 = SSM \, / \, SST \, .$$

This formalizes the interpretation of r^2 as explaining the fraction of variability in the data explained by the regression model.

If $r^2 = 0$, the model does not fit the data. If $r^2 = 1$, the adjustment is perfect. The fair r^2 value to produce an appropriate adjustment is subjective. In the case of exact sciences, $r^2 > 0.9$ are accepted as indicators of a good adjustment. However, regarding social sciences, $r^2 > 0.5$ is acceptable to model's adjustment to the data.

Considering the data of this study, to understand the influence of Gender, Python_user, R_user, Age variables in the number of publications, a linear regression could be used. The first step of this analysis is the elimination of outliers or missing values (by default R deletes observations with missing values; in Python, there are useful functions for this task). Furthermore, regression analysis can be applied to numeric or Boolean data. Thus, Gender, Python_user, and R_user variables must be recoded. Consider "female" as the value 1 and "male" as 0. In Python_user and R_user variables, "yes" is the number 1 and "no" is the value 0. See Table 1 and Table 2.

By analyzing the last lines of the R's output, it is possible to verify that two observations were eliminated because they contain missing values. In Python, the missing values are initially removed. Furthermore, the value of the coefficient of determination is displayed (Multiple R-squared, R^2) and the adjusted coefficient of determination (Adjusted R-squared, R^2_a). The Adjusted R-squared presents a value of approximately 0.41. This value means that 41% of the total variability in publications is explained by the independent variables used in the adjusted linear regression model.

It is also indicated that the test statistic value F (approximately 34) is with $p < 0.0001$. Thus, the null hypothesis should be rejected. In this case, the model is highly significant.

However, it is important to know, if all the studied variables significantly contribute to the linear regression model. Thus, the "coefficient" section is the estimation of each variable's coefficients, the standard error, the value of statistic t-test, and the *p*-value. Looking at the column of *p*-values, it is pos-

Table 1. R language: influence of Gender, Python_user, R_user and Age variables in the number of publications

	In R

Code	`### Influence Gender, Python_user, R_user, Age variables in the number of publications`
	`# Data recoded`
	`data$Gender<-ifelse(data$Gender=="female", 1, 0)`
	`data$Python_user<-ifelse(data$Python_user=="yes", 1, 0)`
	`data$R_user<-ifelse(data$R_user=="yes", 1, 0)`
	`# Multiple Linear Regression for Publications`
	`fit <- lm(Publications ~ Gender + Python_user + R_user + Age, data=data)`
	`# Show results`
	`summary(fit)`

Output	`Call:`
	`lm(formula = Publications ~ Gender + Python_user + R_user + Age, data = data)`
	`Residuals:`
	`Min 1Q Median 3Q Max`
	`-15.3825 -3.8510 -0.1819 4.0661 28.7784`
	`Coefficients:`
	`Estimate Std. Error t value Pr(>|t|)`
	`(Intercept) -4.16855 2.91467 -1.430 0.1543`
	`Gender 1.15397 0.91782 1.257 0.2102`
	`Python_user 0.41310 0.90026 0.459 0.6468`
	`R_user 1.76949 0.90694 1.951 0.0525.`
	`Age 0.86494 0.07568 11.430 <2e-16 ***`
	`---`
	`Signif. codes: 0 '***' 0.001 '**' 0.01 '*' 0.05 '.' 0.1 ' ' 1`
	`Residual standard error: 5.975 on 193 degrees of freedom`
	`(2 observations deleted due to missingness)`
	`Multiple R-squared: 0.4168, Adjusted R-squared: 0.4047`
	`F-statistic: 34.49 on 4 and 193 DF, p-value: < 2.2e-16`

sible to verify that only the variable Age is significant. Besides, the R_user variable has a *p*-value near 0.05. As we have been working with a 5% level of significance, and this variable is very close, it will be considered in this analysis. Then, it can also be concluded that only R_user variables and Age contribute to the explanation of the linear regression model.

Regarding the value of the coefficients of significant variables, it is possible to conclude that:

- R users have more publications than non-users (positive coefficient).
- One year in the data represents an increase in the number of publications, on average, 0.86494.

Table 2. Python language: influence of Gender, Python_user, R_user and Age variables in the number of publications

In Python

Code

```
### Influence Gender, Python_user, R_user, Age variables in the
number of publications
# Import libraries/modules
import numpy as np
import statsmodels.formula.api as smf
# Remove line in data where Age=NaN
datadf = datadf[np.isfinite(datadf['Age'])]
# Remove line in data where Python_user=NaN
datadf = datadf.dropna(subset=['Python_user'])
# Replace values
datadf['Gender'] = datadf['Gender'].replace(['male','fema
le'],[0,1])
datadf['Python_user'] = datadf['Python_user'].
replace(['no','yes'],[0,1])
datadf['R_user'] = datadf['R_user'].replace(['no','yes'],[0,1])
# Fit regression model (using the natural log of one of the
regressors)
results = smf.ols('Publications ~ Gender + Python_user + R_user +
Age', data=datadf).fit()
# Inspect the results
print(results.summary())
```

Output

```
print(results.summary())
OLS Regression Results
==============================================================================
Dep. Variable:          Publications   R-squared:                       0.415
Model:                           OLS   Adj. R-squared:                  0.403
Method:                Least Squares   F-statistic:                     34.11
Date:               Thu, 18 Aug 2016   Prob (F-statistic):           1.70e-21
Time:                       12:42:32   Log-Likelihood:                -629.64
No. Observations:                197   AIC:                             1269.
Df Residuals:                    192   BIC:                             1286.
Df Model:                          4
Covariance Type:           nonrobust
==============================================================================
                    coef    std err         t      P>|t|   [95.0% Conf. Int.]
------------------------------------------------------------------------------
Intercept        -4.2272      2.940    -1.438      0.152    -10.026      1.571
Python_user[T.1]  0.4312      0.908     0.475      0.635     -1.360      2.222
Gender            1.1396      0.924     1.234      0.219     -0.682      2.961
R_user            1.7816      0.912     1.954      0.052     -0.017      3.580
Age               0.8661      0.076    11.376      0.000      0.716      1.016
==============================================================================
Omnibus:                      22.483   Durbin-Watson:                   2.196
Prob(Omnibus):                 0.000   Jarque-Bera (JB):               47.025
Skew:                          0.533   Prob(JB):                     6.15e-11
Kurtosis:                      5.143   Cond. No.                         260.
==============================================================================
Warnings:
[1] Standard Errors assume that the covariance matrix of the errors is
correctly specified.
```

CONCLUSION

Regression analysis can be more exploited than is presented in this book. However, the goal is to give the reader access to the most important points needed to do an optimal data analysis. Consequently, in this chapter, the explained main concepts are:

- ANOVA for regression,
- Coefficient of determination,
- Multiple Linear Regression.

Chapter 7
Factor Analysis

INTRODUCTION

Factor analysis is a statistical method used to describe variability among observed, correlated variables. The goal of performing factor analysis is to search for some unobserved variables called factors. This analysis might lead, for example, to the conclusion that it is possible that three unobserved latent variables are reflected in the variations of seven observed variables. The observed variables are modeled as linear combinations of the possible factors, added the error quantification of this approximation. This added information about the interaction of observed variables could be used for further analysis of the importance of each variable in the context of the dataset.

Factor analysis is used in many areas of statistical analysis like, for example, marketing, social sciences, psychology and other situations where a reduction of a large set of variables is adequate to the study being provided. This way, some observed variables are substituted by a set of latent variables in a lower amount, and that, therefore, represent the data in a summarized fashion.

Factor analysis started by being developed before the appearance of modern computers. This beginning of the method was named exploratory factor analysis (EFA). Other variations of factor analysis (for example, confirmatory factor analysis - CFA) will not be explored in this book. Thus, an example of a factorial analysis is presented below.

DOI: 10.4018/978-1-68318-016-6.ch007

Example of a Factorial Analysis

Imagine a Ph.D. Supervisor wants to test the hypothesis there are two kinds of students. A student that "procrastinates" his studies, and the student that does "not procrastinate", neither of which is an observed variable. Thus, the supervisor only has access to the grades of the student in the several phases a Ph.D. has. Suppose there are ten stages and the student is classified in all those stages. Additionally, the supervisor has a database of 500 Ph.D. students. By choosing each student randomly from this vast universe of students, imagine the grades as being random variables also. The supervisor hypothesis might clarify that for each of the 10 Ph.D. grades, the score averaged over the group of all students who share some common pair of values for procrastination and "not procrastinating" is some constant multiplied by their level of procrastination plus another constant multiplied by their level of low inertia behaviour, i.e., it is a combination of those two "factors".

The numbers for a particular stage, by which the two kinds of behavior are multiplied to obtain the expected score, are posited by the hypothesis to be the same for all procrastination level pairs and are called "factor loading" for this subject. For example, the assumption may hold that the average student's aptitude in the field of "State-of-the-Art writing" is {11 × the student's "procrastinating"} + {5 × the student's "not procrastinating"}.

The numbers 11 and 5 are the factor loadings associated with the task of writing the State-of-the-Art chapter. Other academic tasks may have different factor loadings.

Two students having similar degrees of procrastination and equal degrees of having low inertia may have different aptitudes in State-of-the-Art writing because individual skills differ from average abilities. That difference is called the "error" - a statistical term that means the amount by which an individual changes from what is average for his or her levels of procrastination.

The observable data that go into factor analysis would be ten stage's scores of each of the 500 students, a total of 5,000 numbers. The factor loadings and levels of the two kinds of inertia of each student should be inferred from the data.

THE FACTOR ANALYSIS MODEL

The scores of p population variables, extracted from a population with mean's vector μ and variance-covariance matrix £ , can be modeled by:

$$x_1 = \mu_1 + \lambda_{11}f_1 + \lambda_{12}f_2 + \ldots + \lambda_{1m}f_m + \eta_1$$
$$x_2 = \mu_2 + \lambda_{21}f_1 + \lambda_{22}f_2 + \ldots + \lambda_{2m}f_m + \eta_2$$
$$\ldots$$
$$x_p = \mu_p + \lambda_{p1}f_1 + \lambda_{p2}f_2 + \ldots + \lambda_{pm}f_m + \eta_p$$

where f_m are factor values (with $m < p$), η_p represent the p specific factors and λ_{ij} represents the weight of j factor in the variable i (factor loadings), that is, each λ_{ij} measures the contribution of the j common factor in the variable i. Without loss of generality, and for convenience, x_i variables can be centered and reduced as $z_i = (x_i - \mu_i) / \sigma_i$. Thus, the factor model can be written by:

$$z_i = \lambda_{i1}f_1 + \lambda_{i2}f_2 + \ldots + \lambda_{im}f_m + \eta_i \left(i = 1, \ldots, p\right)$$

Note that λ_{ij} values are different depending on whether the analysis is done with the x_i values (factor weights) or z_i (standardized factor weights). It must, therefore, be assumed that (Maroco, 2011):

- Common factors (f_k) are independent (orthogonal) and equally distributed with mean 0 and variance 1 $\left(k = 1, \ldots, m\right)$.
- Specific factors (η_j) are independent and equally distributed with mean 0 and variance $\psi_j, \left(j = 1, \ldots, p\right)$.
- f_k and η_j are independent.

R VS. PYTHON

An example of factor analysis will be given in this chapter. Consider, for instance, the data from a survey presented to the researchers. The goal of the surveys was to understand the researcher's behavior. The questionnaire presented to the researchers was explained in the chapter "Dataset" of this book. Since different questions seem to evaluate the same characteristic, a division by factor is convenient.

Therefore, the first step is describing the variables, which allows discovering some irregularities such as missing values or outliers. See Table 1 and Table 2.

The output above shows the non-existence of missing values. As the answers are represented on a Likert scale, outliers should not exist, except when there are data errors. In this case, the errors should be corrected, or the particular researcher's data row should be eliminated. As it can be seen, all variables vary between 1 and 5.

Depending on the variable type, different methods to obtain the correlation matrix should be used. In the case of quantitative variables, Pearson correlations have satisfactory results. However, in the case of nominal and ordinal variables, several authors (Marôco, 2011) advocate the use of other types of correlations. This is regard to tetrachoric correlations for nominal variables and polychoric correlations for ordinal data.

Although the polychoric's correlations present an excellent performance, the calculation and validation of the associated assumptions require significant samples with $n > 1000$. This condition limits the use of polychoric correlations for smaller samples. Consequently, a highly clear analysis strategy for qualitative data is the use of Cramer's V correlations for nominal variables, or Spearman correlations for ordinal variables. Regarding that Q1 to Q10 variables are ordinal, in the example presented in this chapter, Spearman's correlation will be the most suitable. See Table 3 and Table 4.

Table 1. R language: descriptive analysis of Q1 to Q10 variables

	In R

Code	`### Descriptive analysis of Q1 to Q10 variables`
	`# Identification of the variables used in factor analysis`
	`survey<-data[, paste("Q", 1:10, sep="")]`
	`# Descriptive analysis for each variable`
	`summary(survey)`

Output	see below

```
       Q1              Q2              Q3              Q4              Q5
 Min.   :1.000   Min.   :2.00   Min.   :1.000   Min.   :1.00   Min.   :1.000
 1st Qu.:3.000   1st Qu.:3.00   1st Qu.:3.000   1st Qu.:3.00   1st Qu.:3.000
 Median :4.000   Median :4.00   Median :4.000   Median :3.00   Median :4.000
 Mean   :3.545   Mean   :3.92   Mean   :3.865   Mean   :3.21   Mean   :3.585
 3rd Qu.:4.000   3rd Qu.:4.25   3rd Qu.:5.000   3rd Qu.:4.00   3rd Qu.:5.000
 Max.   :5.000   Max.   :5.00   Max.   :5.000   Max.   :5.00   Max.   :5.000
       Q6             Q7             Q8            Q9              Q10
 Min.   :1.00   Min.   :1.00   Min.   :1.0   Min.   :1.000   Min.   :1.000
 1st Qu.:3.00   1st Qu.:3.75   1st Qu.:2.0   1st Qu.:3.000   1st Qu.:3.000
 Median :4.00   Median :4.00   Median :3.0   Median :4.000   Median :4.000
 Mean   :3.38   Mean   :4.01   Mean   :3.1   Mean   :3.825   Mean   :3.905
 3rd Qu.:4.00   3rd Qu.:5.00   3rd Qu.:4.0   3rd Qu.:5.000   3rd Qu.:5.000
 Max.   :5.00   Max.   :5.00   Max.   :5.0   Max.   :5.000   Max.   :5.000
```

Table 2. Python language: descriptive analysis of Q1 to Q10 variables

In Python

Code

```
### Descriptive analysis of the variables Q1 to Q10
# Read csv with survey data
import pandas as pd
data = pd.read_csv('newdata.csv', sep=';')
survey_data = data.ix[:,7:17]
# Summary of the variables
survey_data.describe()
```

Output

```
              Q1          Q2         Q3          Q4         Q5          Q6
count  200.000000  200.000000  200.00000  200.000000  200.00000  200.000000
mean     3.545000    3.920000    3.86500    3.210000    3.58500    3.380000
std      1.210496    0.834916    1.03544    0.959428    1.18736    1.082339
min      1.000000    2.000000    1.00000    1.000000    1.00000    1.000000
25%      3.000000    3.000000    3.00000    3.000000    3.00000    3.000000
50%      4.000000    4.000000    4.00000    3.000000    4.00000    4.000000
75%      4.000000    4.250000    5.00000    4.000000    5.00000    4.000000
max      5.000000    5.000000    5.00000    5.000000    5.00000    5.000000
              Q7         Q8          Q9         Q10
count  200.000000  200.00000  200.000000  200.000000
mean     4.010000    3.10000    3.825000    3.905000
std      0.951011    1.13421    1.188037    1.127979
min      1.000000    1.00000    1.000000    1.000000
25%      3.750000    2.00000    3.000000    3.000000
50%      4.000000    3.00000    4.000000    4.000000
75%      5.000000    4.00000    5.000000    5.000000
max      5.000000    5.00000    5.000000    5.000000
```

Table 3. R language: correlation between variables Q1 to Q10

In R

Code

```
### Correlation between variables Q1 to Q10
correlation <- cor(survey, method="spearman")
correlation
```

Output

```
        Q1     Q2     Q3     Q4     Q5     Q6     Q7     Q8     Q9    Q10
Q1  1.0000 0.2217 0.2231 0.2567 0.4320 0.4041 0.2541 0.2718 0.5374 0.2845
Q2  0.2217 1.0000 0.3128 0.2653 0.2390 0.1934 0.1384 0.2253 0.0271 0.1507
Q3  0.2231 0.3128 1.0000 0.1542 0.2133 0.2597 0.2352 0.2377 0.0721 0.2486
Q4  0.2567 0.2653 0.1542 1.0000 0.4236 0.1459 0.0817 0.3166 0.1489 0.2511
Q5  0.4320 0.2390 0.2133 0.4236 1.0000 0.3824 0.1575 0.4381 0.4419 0.2671
Q6  0.4041 0.1934 0.2597 0.1459 0.3824 1.0000 0.2281 0.2218 0.3081 0.3389
Q7  0.2541 0.1384 0.2352 0.0817 0.1575 0.2281 1.0000 0.1956 0.3257 0.2518
Q8  0.2718 0.2253 0.2377 0.3166 0.4381 0.2218 0.1956 1.0000 0.2751 0.2079
Q9  0.5374 0.0271 0.0721 0.1489 0.4419 0.3081 0.3257 0.2751 1.0000 0.2491
Q10 0.2845 0.1507 0.2486 0.2511 0.2671 0.3389 0.2518 0.2079 0.2491 1.0000
```

The outputs above show the matrix of correlations for Q1 to Q10 variables. The correlation values (between different questions) varies from 0.027 (Q2 and Q9) to 0.537 (Q1 and Q9). The data of this matrix and this correlation values suggest the variables could be reduced down to at least two underlying variables or factors. This is a preliminary conclusion, as the correlation

Table 4. Python language: correlation between variables Q1 to Q10

	In Python

Code	`### Correlation between variables Q1 to Q10`
	`survey_data_corr = survey_data.corr(method='spearman')`
	`survey_data_corr`

Output	

	Q1	Q2	Q3	Q4	Q5	Q6	Q7
Q1	1.000000	0.221737	0.223101	0.256695	0.432009	0.404096	0.254117
Q2	0.221737	1.000000	0.312831	0.265311	0.239049	0.193397	0.138398
Q3	0.223101	0.312831	1.000000	0.154184	0.213342	0.259665	0.235240
Q4	0.256695	0.265311	0.154184	1.000000	0.423595	0.145878	0.081694
Q5	0.432009	0.239049	0.213342	0.423595	1.000000	0.382424	0.157458
Q6	0.404096	0.193397	0.259665	0.145878	0.382424	1.000000	0.228059
Q7	0.254117	0.138398	0.235240	0.081694	0.157458	0.228059	1.000000
Q8	0.271804	0.225324	0.237712	0.316570	0.438128	0.221814	0.195582
Q9	0.537406	0.027119	0.072103	0.148947	0.441893	0.308125	0.325718
Q10	0.284481	0.150652	0.248648	0.251149	0.267063	0.338883	0.251794

	Q8	Q9	Q10
Q1	0.271804	0.537406	0.284481
Q2	0.225324	0.027119	0.150652
Q3	0.237712	0.072103	0.248648
Q4	0.316570	0.148947	0.251149
Q5	0.438128	0.441893	0.267063
Q6	0.221814	0.308125	0.338883
Q7	0.195582	0.325718	0.251794
Q8	1.000000	0.275094	0.207895
Q9	0.275094	1.000000	0.249120
Q10	0.207895	0.249120	1.000000

values vary from very little correlation to moderate correlation between the pairs of survey questions. Nonetheless, we can have more factors, and additional research should be addressed with this data.

In the following subsections, we present some suggestions to do factor analysis, in R and Python.

Sampling Adequacy

Before starting factor analysis, it should be checked whether it is appropriate to the data in the study. For this verification, two methods could be applied: the Bartlett sphericity test and the KMO Measure.

Bartlett Sphericity Test

Exploratory factor analysis is only useful if the matrix of population correlation is statistically different from the identity matrix. If these are equal, the variables are few interrelated, i.e., the specific factors explain the greater proportion of the variance and the common factors are unimportant. Therefore, it should be defined when the correlations between the original variables are sufficiently high. Thus, the factor analysis is useful in estimation of common factors. With this in mind, the Bartlett Sphericity test can be used. The hypotheses are:

H_0: The matrix of population correlations is equal to the identity matrix.
H_1: The matrix of population correlations is different from the identity matrix.

This test can be calculated in R or Python. However, with some packages, it can be made more quickly. It is the case of the *PSYCH* R's package, which proposes Bartlett's test. In Python, there is a need to program the Bartlett Sphericity test function. To our best knowledge, there is no implementation of this test in any free module. Thus, Bartlett's test statistic formula is:

$$X^2 = -\left(n - 1 - \frac{2p+5}{6}\right) \times \ln|R|$$

where $|R|$ is the determinant of the correlation matrix, and $\left[p \times \frac{(p-1)}{2}\right]$ is the number of degrees of freedom with p as the number of variables. The programming code for this calculation is shown in Table 5 and Table 6.

With the previous results, it is possible to conclude the rejection of the null hypothesis ($p<0.05$), and therefore, the matrix of population correlations is different from the identity matrix. This difference suggests that factorial analysis is appropriate to our data.

However, Bartlett Sphericity test is rarely used since it is very sensitive to the dimension of the sample. In large samples, H_0 is frequently rejected even

Table 5. R language: Bartlett Sphericity test

	In R

Code	`### Bartlett Sphericity test`
	`library (psych)`
	`cortest.bartlett(correlation, n=nrow(data))`

Output	`$chisq`
	`[1] 410.2728`
	`$p.value`
	`[1] 1.949995e-60`
	`$df`
	`[1] 45`

Table 6. Python language: Bartlett Sphericity test

	In Python

Code	`### Bartlett Sphericity test`
	`import numpy as np`
	`import math as math`
	`import scipy.stats as stats`
	`# Generate Identity Matrix`
	`# 10x10 Identity Matrix`
	`indentity = np.identity(10)`
	`# The Bartlett test`
	`n = survey_data.shape[0]`
	`p = survey_data.shape[1]`
	`chi2 = -(n-1-(2*p+5)/6)*math.log(np.linalg.det(survey_data_corr))`
	`ddl = p*(p-1)/2`
	`pvalue = stats.chi2.pdf(chi2, ddl)`
	`chi2`
	`ddl`
	`pvalue`

Output	`chi2`
	`Out[12]: 410.27280642443156`
	`ddl`
	`Out[13]: 45.0`
	`pvalue`
	`Out[14]: 8.7335941050291506e-61`

when the correlations are very small. The test also requires the multivariate normal distribution of the variables, and it is very sensitive to this assumption's violation.

KMO Measure

Given the limitations of Bartlett Sphericity test, there are other methods with the same goal that can be used to assess the quality of data. A widely used method is the "measure of the adequacy of sampling Kaiser-Meyer-Olkin" (KMO). KMO checks if it is possible to factorize the primary variables efficiently. But it is based on another idea.

The correlation matrix is always the starting point. The variables are more or less correlated, but the others can influence the correlation between the two variables. Hence, with KMO, the partial correlation is used to measure the relation between two variables by removing the effect of the remaining variables.

The partial correlation matrix can be obtained from the correlation matrix. Considering the inverse of the correlation matrix as $R^{-1} = \left(v_{ij}\right)$, the partial correlation as $A = \left(a_{ij}\right)$, and the observed correlation matrix as $R = \left(r_{ij}\right)$, we have:

$$a_{ij} = -\frac{v_{ij}}{\sqrt{v_{ii} \times v_{jj}}}$$

Thus, the overall KMO index is computed as:

$$KMO = \frac{\sum_i \sum_{j \neq i} r_{ij}^2}{\sum_i \sum_{j \neq i} r_{ij}^2 + \sum_i \sum_{j \neq i} a_{ij}^2}$$

and the KMO index per variable to detect those which are not related to the others is:

$$KMO_j = \frac{\sum_{j \neq i} r_{ij}^2}{\sum_{j \neq i} r_{ij}^2 + \sum_{j \neq i} a_{ij}^2}$$

KMO returns values between 0 and 1. A rule of thumb for interpreting the statistic:

- KMO values between 0.8 and 1 indicate the sampling is adequate.
- KMO values less than 0.6 indicate the sampling is not appropriate and that remedial action should be taken. Some authors put this value at 0.5, so the researcher should use his judgment for values between 0.5 and 0.6.
- KMO Values close to zero means that there are high partial correlations compared to the sum of correlations. In other words, there are widespread correlations, which are a large problem for factor analysis.

For reference, Kaiser suggested the following classification of the results:

- 0 to 0.49 unacceptable.
- 0.50 to 0.59 miserable.
- 0.60 to 0.69 mediocre.
- 0.70 to 0.79 middling.
- 0.80 to 0.89 meritorious.
- 0.90 to 1.00 marvelous.

Similar to Bartlett Sphericity test, the KMO measure could be calculated in R or Python (Table 7 and Table 8). However, with the package used above (PSYCH), in R, the code is much simpler. In Python, this test will be programmed since, to our best knowledge, a module with KMO was not available.

With KMO = 0.8, the degree of common variance in our dataset is "meritorious". All variables have KMO higher than 0.5, and therefore, the factor analysis is appropriate to this data. However, if KMO is less than 0.5, a detailed discussion of the variables should be made.

Retained Factors

Since it is possible to make a factor analysis to the data in the study, the next step is to find the weights for a set of latent factors. However, this type of mathematical model has multiple possible solutions. This problem is referred to as indeterminacy of Exploratory Factor Analysis (EFA) equation caused by the problem of factors rotation. Therefore, whenever a solution is not interpretable, it can be made a rotation of factors (multiplication by an orthogonal matrix). The "rotation" is equivalent to the translation of factorial axes in the factorial space without changing the orientation of the vectors representing the variables.

Number of Factors to Be Retained

Before starting to analyze factors, it is important to know how many factors should be maintained. Several studies were developed to decide this number. Hayton et al. (2004) state three reasons why this decision is so important. Firstly, it can affect EFA results more than other decisions, such as selecting an extraction method or the factor rotation method, since there is evidence of the relative robustness of EFA with regards to these matters. Secondly,

Table 7. R language: KMO Measure

	In R

Code	`### KMO Measure`
	`library (psych)`
	`KMO(correlation)`

Output	`Kaiser-Meyer-Olkin factor adequacy`
	`Call: KMO(r = correlation)`
	`Overall MSA = 0.8`
	`MSA for each item =`
	`  Q1   Q2   Q3   Q4   Q5   Q6   Q7   Q8   Q9  Q10`
	`0.81 0.77 0.79 0.77 0.80 0.84 0.79 0.86 0.71 0.86`

Table 8. Python language: KMO Measure

	In Python

Code	```### KMO Measure
	import numpy as np
	import math as math
	
	### a) Global KMO
	# Inverse of the correlation matrix
	# dataset_corr is the correlation matrix of the survey results
	corr_inv = np.linalg.inv(dataset_corr)
	# number of rows and number of columns
	nrow_inv_corr, ncol_inv_corr = dataset_corr.shape
	# Partial correlation matrix
	A = np.ones((nrow_inv_corr,ncol_inv_corr))
	for i in range(0,nrow_inv_corr,1):
	for j in range(i,ncol_inv_corr,1):
	#above the diagonal
	A[i,j] = - (corr_inv[i,j])/(math.sqrt(corr_inv[i,i] * corr_inv[j,j]))
	#below the diagonal
	A[j,i] = A[i,j]
	#transform to an array of arrays ("matrix" with Python)
	dataset_corr = np.asarray(dataset_corr)
	#KMO value
	kmo_num = np.sum(np.square(dataset_corr))-np.sum(np.square(np.
	diagonal(dataset_corr)))
	kmo_denom = kmo_num + np.sum(np.square(A)) - np.sum(np.square(np.
	diagonal(A)))
	kmo_value = kmo_num / kmo_denom
	print(kmo_value)
	
	### b) KMO per variable
	#creation of an empty vector to store the results per variable. The
	size of the vector is equal to the number #...of variables
	kmo_j = [None]*dataset_corr.shape[1]
	for j in range(0, dataset_corr.shape[1]):
	kmo_j_num = np.sum(dataset_corr[:,[j]] ** 2) - dataset_corr[j,j] ** 2
	kmo_j_denom = kmo_j_num + np.sum(A[:,[j]] ** 2) - A[j,j] ** 2
	kmo_j[j] = kmo_j_num / kmo_j_denom
	print(kmo_j)```

Output	```a) Global KMO
	    ...: print(kmo_value)
	    0.798844102413
	
	b) KMO per item
	    ...: print(kmo_j)
	    0.812160468405 0.774161264483 0.786819432663 0.766251123086
	0.800579196084 0.842927745203 0.792010173432 0.862037322891
	0.714795031915 0.856497242574```

the EFA requires that balance is struck between "reducing" and adequately "representing" the correlations that exist in a group of variables. Therefore, its very usefulness depends on distinguishing relevant factors from trivial ones. Lastly, an error regarding selecting the number of factors can significantly alter the solution and the interpretation of EFA results. The extraction of fewer factors can lead to the loss of relevant information and a substantial distortion in the solution (for example, in the loading variables). On the other hand, although less problematic, the extraction of an excessive number of factors can lead to factors with a substantial less loading. Thus, it can be difficult to interpret and/or replicate.

Given the importance of this decision, different methods have been proposed to determine the number of factors to retain.

Kaiser Criterion

This is a method suggested by Kaiser (1960). According to his rule, only factors with eigenvalues greater than one are retained for interpretation. Despite the simplicity of this approach, many authors agree that it is problematic and inefficient when it comes to determining the number of factors (Ledesma and Valero-Mora, 2007). For example, it doesn't make much sense to regard a factor with an eigenvalue of 1.01 as "major" and one with an eigenvalue of.99 as "trivial". This method should be used together with other methods. See Table 9 and Table 10.

The outputs are divided into "values" and "vectors", corresponding to eigenvalues and eigenvectors, respectively. As it is possible to observe, there are three eigenvalues higher than 1. Please note that, in Python, eigenvalues are not ordered. This suggests retaining three factors.

Scree Plot

This is a method proposed by Cattell (1966), which involves the visual exploration of a graphical representation of the eigenvalues. In this approach, the eigenvalues are presented in descending order and linked with a line. Afterward, the graph is examined to determine the point at which the last significant drop or break takes place - in other words, where the line levels off. The logic behind this method is that the point divides the critical or major factors from the minor or unimportant factors. See Table 11 and Table 12. Scree plot has been criticized for its subjectivity since there is not an objective definition of the cutoff point between the important and trivial factors. Indeed, some cases may present several drops and possible cutoff points, such that the graph may be ambiguous and difficult to interpret.

The observation of the previous scree plots suggests that at least two factors should be retained. In this point, the curve makes an "elbow" toward less steep decline. Other suggestion is to consider four factors (little elbow).

Table 9. R language: Kaiser criterion

	In R

Code	`### Kaiser criterion`
	`library (psych)`
	`eigen(correlation)`

Output	see below

```
$values
[1] 3.3669166 1.2136498 1.0776487 0.8213321 0.7951511 0.7124396 0.6119054
0.5670223
[9] 0.4679005 0.3660340
$vectors
           [,1]         [,2]        [,3]        [,4]        [,5]        [,6]
 [1,] -0.3853256  0.256021260  0.04702543  0.05434129 -0.34124508 -0.24403301
 [2,] -0.2385566 -0.554897142 -0.12377747 -0.10793243 -0.37755610 -0.46948278
 [3,] -0.2590495 -0.384579932 -0.47012641 -0.05655833 -0.10685503  0.37008699
 [4,] -0.2827407 -0.338060670  0.45012288  0.08480756  0.36816709 -0.33303133
 [5,] -0.3970827  0.003995701  0.37177858  0.02653913 -0.09457523  0.14696031
 [6,] -0.3376002  0.129020624 -0.20801818  0.46735905 -0.29588083  0.23702543
 [7,] -0.2552794  0.207102791 -0.48687588 -0.53430300  0.33095236 -0.25730614
 [8,] -0.3229872 -0.162204890  0.26236628 -0.41518560  0.15971844  0.56139937
 [9,] -0.3385482  0.527492060  0.11965551 -0.17450361 -0.09880217 -0.10303279
[10,] -0.3030087  0.015600699 -0.23905186  0.51726913  0.59392502 -0.04888472
             [,7]       [,8]         [,9]       [,10]
 [1,] -0.31897190  0.1916059  0.514234180 -0.44935505
 [2,]  0.32625111  0.3132422 -0.159056263  0.12404674
 [3,] -0.61538208 -0.1078221 -0.082569260  0.12350375
 [4,] -0.19358060 -0.4319708  0.251440196  0.24943628
 [5,]  0.01215364 -0.1942442 -0.650994313 -0.45953434
 [6,]  0.46031902 -0.4160287  0.190036142  0.20705075
 [7,]  0.20849609 -0.3364610 -0.009101036 -0.19213969
 [8,]  0.29342881  0.3084268  0.324548292  0.03025459
 [9,] -0.18018807  0.1977266 -0.250906346  0.63833622
[10,]  0.03960368  0.4568397 -0.109310158 -0.07663455
```

Variance Explained Criteria

This is a method based on similar conceptual structure is to retain the number of factors that account for a certain percent of extracted variance. The literature varies on how much variance should be explained before the number of factors is sufficient. The majority suggests that 75-90% of the variance should be accounted. However, some statisticians indicate as much as 50% of the variance explained is acceptable. As with any criteria method solely depending on variance, this seemingly full standard must be viewed about to the foundational differences between extraction methods. See Table 13 and Table 14.

The previous outputs give the importance of each component for the number of variables in the study, i.e., 10. To have 75% of the explained variance, it would be necessary to consider six components. However, for example, in psychology or sociology, these values are difficult to reach with a small number of components. In this case, the minimum acceptable value

Table 10. Python language: Kaiser criterion

	In Python
Code	`### Kaiser criterion` `np.linalg.eig(survey_data_corr)`

Output	`np.linalg.eig(survey_data_corr)`
	`Out[20]:`
	`(array([ 3.36691657, 1.21364977, 0.366034, 1.0776487, 0.46790045,`
	`0.56702233, 0.61190541, 0.71243962, 0.79515105, 0.8213321 ]),`
	`matrix([[ 0.38532555, 0.25602126, 0.44935505, -0.04702543, -0.51423418,`
	`0.19160593, -0.3189719, 0.24403301, 0.34124508, 0.05434129],`
	`[ 0.2385566, -0.55489714, -0.12404674, 0.12377747, 0.15905626,`
	`0.31324219, 0.32625111, 0.46948278, 0.3775561, -0.10793243],`
	`[ 0.25904952, -0.38457993, -0.12350375, 0.47012641, 0.08256926,`
	`-0.10782206, -0.61538208, -0.37008699, 0.10685503, -0.05655833],`
	`[ 0.28274072, -0.33806067, -0.24943628, -0.45012288, -0.2514402,`
	`-0.43197081, -0.1935806, 0.33303133, -0.36816709, 0.08480756],`
	`[ 0.39708267, 0.0039957, 0.45953434, -0.37177858, 0.65099431,`
	`-0.19424424, 0.01215364, -0.14696031, 0.09457523, 0.02653913],`
	`[ 0.33760022, 0.12902062, -0.20705075, 0.20801818, -0.19003614,`
	`-0.41602867, 0.46031902, -0.23702543, 0.29588083, 0.46735905],`
	`[ 0.25527937, 0.20710279, 0.19213969, 0.48687588, 0.00910104,`
	`-0.33646105, 0.20849609, 0.25730614, -0.33095236, -0.534303 ],`
	`[ 0.32298719, -0.16220489, -0.03025459, -0.26236628, -0.32454829,`
	`0.30842676, 0.29342881, -0.56139937, -0.15971844, -0.4151856 ],`
	`[ 0.33854818, 0.52749206, -0.63833622, -0.11965551, 0.25090635,`
	`0.1977266, -0.18018807, 0.10303279, 0.09880217, -0.17450361],`
	`[ 0.30300873, 0.0156007, 0.07663455, 0.23905186, 0.10931016,`
	`0.45683972, 0.03960368, 0.04888472, -0.59392502, 0.51726913]]))`

Table 11. R language: Scree plot criterion

	In R

Code	`### Scree plot criterion`
	`library(nFactors)`
	`scree(correlation, hline=-1) # hline=-1 draw a horizontal line at -1`

Output	Scree plot in R: 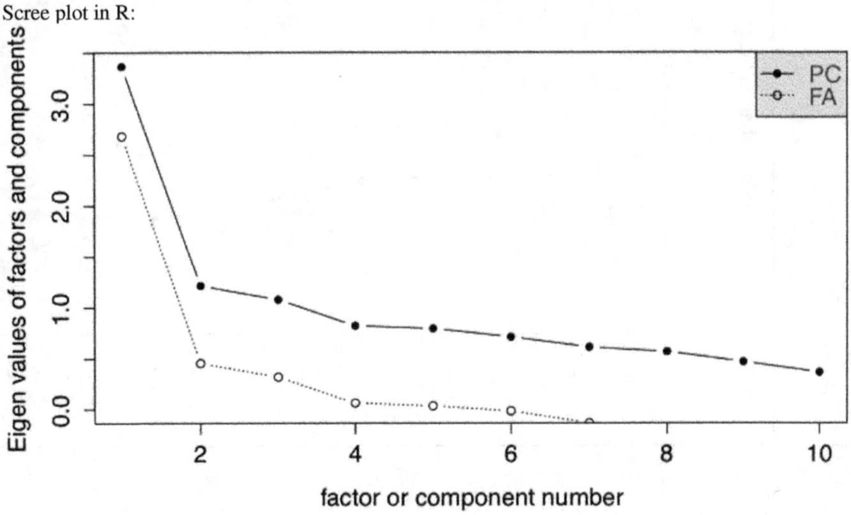

Table 12. Python language: Scree plot criterion

	In Python

Code	```### Scree plot criterion (calling R functions from Python)```
	```# See https://sites.google.com/site/aslugsguidetopython/data-analysis/pandas/calling-r-from-python```
	```# See http://www.lfd.uci.edu/~gohlke/pythonlibs/#rpy2 to install rpy2```
	```import rpy2 as rpy2```
	```from rpy2.robjects.packages import importr```
	```import rpy2.robjects as ro```
	```import pandas.rpy.common as com```
	```# Changing R's directory to where is the data```
	```ro.r('setwd("C:/Users/Rui Sarmento/Documents/Livro Cybertech/Dados e Code")')```
	```# Reading the data with R```
	```ro.r('data_df <- read.csv("data.csv",sep=";")')```
	```# Retrieving the correlation matrix of the survey answers```
	```ro.r('correlation <- cor(data_df[,paste("Q",1:10,sep="")], method="spearman")')```
	```### Scree plot criterion```
	```psych = importr('psych')```
	```# Scree function call with R```
	```ro.r('scree(correlation, hline=-1)') # hline=-1 draw a horizontal line at -1')```
	```library(nFactors)```
	```scree(correlation, hline=-1) # hline=-1 draw a horizontal line at -1```

Output	Scree plot in Python: 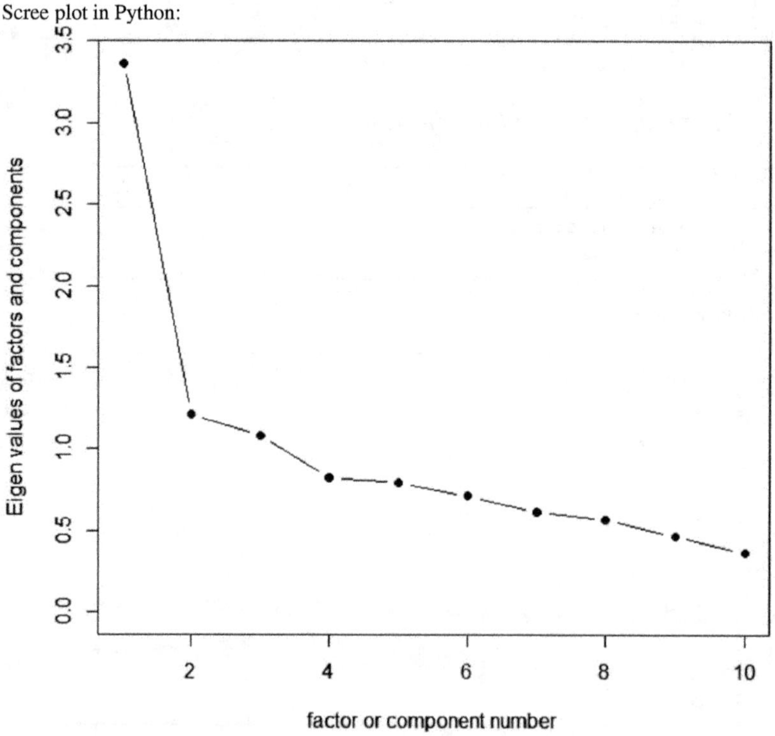

Table 13. R language: explained variance for each component

	In R

Code	`### Explained variance for each component`
	`pc <- prcomp(survey,scale.=F)`
	`summary(pc)`

Output	(see below)

```
Importance of components:
                    PC1    PC2    PC3     PC4     PC5     PC6     PC7     PC8
Standard dev.    2.0616 1.1715 1.0936 0.97953 0.90417 0.89021 0.82018 0.78209
PoV              0.3661 0.1182 0.1030 0.08264 0.07041 0.06825 0.05794 0.05268
Cum. Prop.       0.3661 0.4843 0.5873 0.66990 0.74031 0.80856 0.86650 0.91918
                           PC9    PC10
Standard deviation     0.69473 0.67509
Proportion of Variance 0.04157 0.03925
Cumulative Proportion  0.96075 1.00000
```

Table 14. Python language: explained variance for each component

	In Python

Code	`### Explained Variance for each component`
	`# See http://www.dummies.com/how-to/content/data-science-using-python-to-perform-factor-and-pr.html`
	`from sklearn.decomposition import PCA`
	`import pandas as pd`
	`pca = PCA().fit(survey_data)`
	`pca.explained_variance_ratio_`

Output	`Out[22]:` `array([ 0.36606121, 0.11819385, 0.10300932, 0.08263505, 0.07041005, 0.06825262, 0.05793681, 0.05268024, 0.04156898, 0.03925188])`

(50%) should be considered. Thus, in our case study, three factors should be retained.

To analyze three methods discussed above, in the following analysis, three factors will be considered. However, the reader must have critical thought and check if the number of suggested factors makes sense in the scope of the problem that is being analyzed.

To estimate the matrix of factor weights, it is necessary to have an estimate of the communalities. Among the various methods for this estimation, the most popular are Principal Component Analysis, Principal Axis, and Maximum Likelihood Factor Analysis.

Principal Component Method

The principal component method is based on the determination of the eigenvalues and eigenvectors of the correlation matrix. First, an initial estimate is provided, which is the maximum value of the correlation (i.e., 1). Subsequently, the number of principal components to retain is determined.

Principal Axis Method of Factor Extraction

It is an iterative PCA application to the matrix where communalities stand on the diagonal in place of 1's. Each iteration refines communalities further until they converge. In doing so, the method seeks to explain variance, not pairwise correlations. Principal Axis method has the advantage in that it can, like PCA, analyze not only correlations but also covariance.

Maximum Likelihood Method (ML)

Assumes that correlation came from a population having multivariate normal distribution (other methods make no such assumption) and hence the residuals of correlation coefficients must be normally distributed around 0. The loadings are iteratively estimated by ML Communalities approach under the above assumption. The treatment of correlations is weighted by uniqueness values. While other methods just analyze the sample as it is, ML method allows some inference about the population, some fit indices, and confidence intervals are usually computed along with it.

Succinctly, in most cases, both Principal Component method and Principal Axis method, lead to the same factor structure, and the difference between the methods is mostly conceptual. The Principal Component is the most commonly used method. However, the Principal Axis method is conceptually more attractive, since it assumes a factor structure composed of common factors and specific factors. Consequently, with this method, it is possible to obtain higher factor weights, which facilitates the interpretation of factors. This occurs because they do not have to include the specificity of each variable during the extraction of factors. However, this method is more affected by the indeterminacy of the factors and may cause obtaining very different factor structures from the original data. This disadvantage of the method is particularly penalizing in EFA since it is heavily dependent on the performed sampling. Finally, the maximum likelihood method requires that variables under study present multivariate normal distribution, which is not always easy to validate. This is a reason why the Principal Component method is somewhat recommended. However, it has the advantage that, for large samples, allows the calculation of indices to evaluate the quality of the factor model.

Therefore, the most common method is used in this book: Principal Component method. See Table 15 and Table 16.

Table 15. R language: Principal Component method

	In R

Code	`### Principal Component method`
	`library (psych)`
	`principal(correlation,nfactors=3, rotate="none")`

Output	`Principal Components Analysis`
	`Call: principal(r = correlation, nfactors = 3, rotate = "none")`
	`Standardized loadings (pattern matrix) based upon correlation matrix`
	`     PC1   PC2   PC3   h2   u2 com`
	`Q1  0.71 -0.28 -0.05 0.58 0.42 1.3`
	`Q2  0.44  0.61  0.13 0.58 0.42 1.9`
	`Q3  0.48  0.42  0.49 0.64 0.36 3.0`
	`Q4  0.52  0.37 -0.47 0.63 0.37 2.8`
	`Q5  0.73 -0.01 -0.39 0.68 0.32 1.5`
	`Q6  0.62 -0.14  0.22 0.45 0.55 1.4`
	`Q7  0.47 -0.23  0.50 0.53 0.47 2.4`
	`Q8  0.59  0.18 -0.27 0.46 0.54 1.6`
	`Q9  0.62 -0.58 -0.12 0.74 0.26 2.1`
	`Q10 0.56 -0.02  0.25 0.37 0.63 1.4`
	
	`                       PC1  PC2  PC3`
	`SS loadings           3.37 1.21 1.08`
	`Proportion Var        0.34 0.12 0.11`
	`Cumulative Var        0.34 0.46 0.57`
	`Proportion Explained  0.60 0.21 0.19`
	`Cumulative Proportion 0.60 0.81 1.00`
	`Mean item complexity = 1.9`
	`Test of the hypothesis that 3 components are sufficient.`
	`The root mean square of the residuals (RMSR) is 0.1`
	`Fit based upon off diagonal values = 0.87`

Looking at standardized loadings (pattern matrix) based upon correlation matrix, in the previous output, the first three columns indicate the variable's weight in each defined component. The weights allow defining the variables belonging to each component. Furthermore, h2 and u2 are the values of communalities and uniquenesses, respectively. Communality is the proportion of each variable's variance that can be explained by the factors. Uniqueness gives the percentage of the common variance of the variable not associated with the factors.

$$Uniqueness = 1 - communality$$

Analyzing the communality values, it is possible to verify that all values are higher than 0.5, except Q6, Q8, and Q10. This means that except these three indicated variables, the percentage of the variance of each variable explained by common factors is greater than 50%. Thus, with some statistical rigor, these three variables should be eliminated. However, it is up to the reader to

Table 16. Python language: Principal Component method (using R)

<table>
<tr><th></th><th>In Python</th></tr>
<tr><td>Code</td><td>

```
### Principal Component method (using R)
import rpy2 as rpy2
from rpy2.robjects.packages import importr
import rpy2.robjects as ro
import pandas.rpy.common as com
# Changing R's directory to where data is
ro.r('setwd("C:/Users/Rui Sarmento/Documents/Livro Cybertech/Dados e
Code")')
# Reading data with R
ro.r('data_df <- read.csv("data.csv",sep=";")')
# Retrieving the correlation matrix of the survey answers
ro.r('correlation <- cor(data_df[,paste("Q",1:10,sep="")],
method="spearman")')
# Uses of psych R's package
ro.r('library (psych)')
# Calling function principal with R
print(ro.r('principal(correlation,nfactors=3, rotate="none")'))
library (psych)
principal(correlation,nfactors=3, rotate="none")
```

</td></tr>
<tr><td>Output</td><td>

```
Principal Components Analysis
Call: principal(r = correlation, nfactors = 3, rotate = "none")
Standardized loadings (pattern matrix) based upon correlation matrix
     PC1   PC2   PC3   h2   u2 com
Q1  0.71 -0.28 -0.05 0.58 0.42 1.3
Q2  0.44  0.61  0.13 0.58 0.42 1.9
Q3  0.48  0.42  0.49 0.64 0.36 3.0
Q4  0.52  0.37 -0.47 0.63 0.37 2.8
Q5  0.73 -0.01 -0.39 0.68 0.32 1.5
Q6  0.62 -0.14  0.22 0.45 0.55 1.4
Q7  0.47 -0.23  0.50 0.53 0.47 2.4
Q8  0.59  0.18 -0.27 0.46 0.54 1.6
Q9  0.62 -0.58 -0.12 0.74 0.26 2.1
Q10 0.56 -0.02  0.25 0.37 0.63 1.4
                       PC1  PC2  PC3
SS loadings           3.37 1.21 1.08
Proportion Var        0.34 0.12 0.11
Cumulative Var        0.34 0.46 0.57
Proportion Explained  0.60 0.21 0.19
Cumulative Proportion 0.60 0.81 1.00
Mean item complexity = 1.9
Test of the hypothesis that 3 components are sufficient.
The root mean square of the residuals (RMSR) is 0.1
Fit based upon off diagonal values = 0.87
```

</td></tr>
</table>

decide on this rigor, that is, to determine what is an acceptable lower limit. We should point out that this value should not be less than 30%. Analyzing the weights of each variable in the factors, the reader should check the factor with greater weight for the variable. The variable should belong to this factor. Nevertheless, in the case of Q3, the weights in PC1, PC2, and PC3 are 0.48, 0.42, and 0.49, respectively. This may lead to some doubts regarding the model because there is no sense to decide where the Q3 variable should be included. To eliminate these doubts, the results should be analyzed after a factor rotation (next subsection).

Check the same results considering the elimination of the variables whose communality values are less than 0.5. The results remain not interpretable. See Table 17 and Table 18.

Factor Rotations

EFA solution is not always interpretable. The factor weights of the variables in common factors can be such that it is not possible to assign a meaning to extracted empirical factors. From the mathematical point of view, the extracted factors are not the only existing ones, and an orthogonal matrix can be multiplied by the matrix of factor weights. The multiplication corresponds to the rotation of the factorial axes and does not alter the communalities or the specific variance, i.e., does not modify the data structure.

The factorial axes are mathematical structures and not laws of nature. Hence, there is no reason for an axis system to be preferred over another

Table 17. R language: Principal Component method (check the same results)

	In R

Code	`### Principal Component method`
	`new.correlation<-correlation[!(colnames(correlation) %in% c("Q6", "Q8", "Q10")),!(rownames(correlation) %in% c("Q6", "Q8", "Q10"))]`
	`new.correlation`
	`library (psych)`
	`principal(new.correlation,nfactors=3, rotate="none")`

Output	`Principal Components Analysis`
	`Call: principal(r = new.correlation, nfactors = 3, rotate = "none")`
	`Standardized loadings (pattern matrix) based upon correlation matrix`
	`    PC1   PC2   PC3   h2   u2 com`
	`Q1 0.75 -0.25 -0.02 0.63 0.37 1.2`
	`Q2 0.47  0.65  0.06 0.64 0.36 1.8`
	`Q3 0.47  0.50  0.45 0.67 0.33 3.0`
	`Q4 0.55  0.30 -0.53 0.67 0.33 2.5`
	`Q5 0.74 -0.05 -0.36 0.68 0.32 1.4`
	`Q7 0.48 -0.17  0.65 0.68 0.32 2.0`
	`Q9 0.66 -0.59  0.02 0.79 0.21 2.0`
	`                       PC1  PC2  PC3`
	`SS loadings           2.53 1.20 1.03`
	`Proportion Var        0.36 0.17 0.15`
	`Cumulative Var        0.36 0.53 0.68`
	`Proportion Explained  0.53 0.25 0.22`
	`Cumulative Proportion 0.53 0.78 1.00`
	`Mean item complexity = 2`
	`Test of the hypothesis that 3 components are sufficient.`
	`The root mean square of the residuals (RMSR) is 0.12`
	`Fit based upon off diagonal values = 0.82`

Table 18. Python language: Principal Component method (check the same results)

	In Python

Code	`### Principal Component method`
	`# New correlation matrix`
	`ro.r('new.correlation<-correlation[!(colnames(correlation) %in% c("Q6", "Q8", "Q10")),!(rownames(correlation) %in% c("Q6", "Q8", "Q10"))]')`
	`ro.r('library(psych)')`
	`# Calling function principal with R`
	`print(ro.r('principal(new.correlation,nfactors=3, rotate="none")'))`

Output	`Principal Components Analysis`
	`Call: principal(r = new.correlation, nfactors = 3, rotate = "none")`
	`Standardized loadings (pattern matrix) based upon correlation matrix`
	`   PC1   PC2   PC3   h2   u2 com`
	`Q1 0.75 -0.25 -0.02 0.63 0.37 1.2`
	`Q2 0.47  0.65  0.06 0.64 0.36 1.8`
	`Q3 0.47  0.50  0.45 0.67 0.33 3.0`
	`Q4 0.55  0.30 -0.53 0.67 0.33 2.5`
	`Q5 0.74 -0.05 -0.36 0.68 0.32 1.4`
	`Q7 0.48 -0.17  0.65 0.68 0.32 2.0`
	`Q9 0.66 -0.59  0.02 0.79 0.21 2.0`
	`                       PC1  PC2  PC3`
	`SS loadings            2.53 1.20 1.03`
	`Proportion Var         0.36 0.17 0.15`
	`Cumulative Var         0.36 0.53 0.68`
	`Proportion Explained   0.53 0.25 0.22`
	`Cumulative Proportion  0.53 0.78 1.00`
	`Mean item complexity = 2`
	`Test of the hypothesis that 3 components are sufficient.`
	`The root mean square of the residuals (RMSR) is 0.12`
	`Fit based upon off diagonal values = 0.82`

axis system. Moreover, the best axis system is one that produces a factor solution easily interpretable. There are several methods to make the rotation of the factorial axes, including the Varimax method, the Quartimax method and the Oblimin method.

Varimax

Varimax, which was developed by Kaiser (1958), is indubitably the most popular rotation method by far. For Varimax, a simple solution means that each factor has a small number of large loadings and a large number of zero (or low) loadings. This simplifies the interpretation because, after a varimax rotation, each original variable tends to be associated with one (or a small number) of factors, and each factor represents only a limited number of variables. Additionally, the factors can often be interpreted from the opposition of few variables with positive loadings to few variables with negative loadings (Abdi, 2003).

Quartimax

Quartimax rotation is a form of orthogonal rotation used to transform vectors associated with principal component analysis or factor analysis to a simple structure. It is a particular case of orthomax rotation, which maximizes the sums of squares of the coefficients across the resultant vectors for each of the original variables, Quartimax is opposed to varimax, which maximizes the sums of squares of the coefficients within each of the resultant vectors.

Oblimin

Oblimin rotation is a general form for obtaining oblique rotations used to transform vectors associated with principal component analysis or factor analysis to a simple structure. Oblimin is similar to the Orthomax rotation procedures used in orthogonal rotation in that, it too, includes an arbitrary constant used to obtain different rotational properties. While most orthogonal rotations use some form of Orthomax rotation, this is no longer the case with Oblimin rotation for the oblique case.

For the next step, the varimax method was used. The results are presented in Table 19 and Table 20.

In this case, PC3 is composed of Q1, Q6, Q7, Q9, and Q10 variables. PC1 is composed of Q4, Q5, and Q8 variables. PC2 is composed of Q2 and Q3 variables.

The RMSR value equal to 0.1 means that the retained factors are appropriate to describe the correlation structure. As reference values, the model is considered unacceptable when RMSR value is greater than 0.1. It is considered as excellent when RMSR value is less than 0.05.

Check the same results considering the elimination of the variables whose communality values are less than 0.5. See Table 21 and Table 22.

The previous outputs show that communality values are always greater than 60%. Moreover, PC3 is composed of Q1, Q7, and Q9 variables. PC1 consists of Q4 and Q5 variables. PC2 is composed of Q2 and Q3 variables. In this case, RMSR is equal to 0.12, and then the model is unacceptable.

Remembering the "Dataset" chapter, we can clearly relate each factor to a subject of study in the questionnaire. Therefore, the first factor groups the questions related to research tools, the second factor groups the questions related to research methods. Finally, the third factor groups the productivity related questions. Consequently, in the study of this book, all variables must be considered, and the factors are:

Table 19. R language: Principal Component method with varimax rotation

	In R

Code	`### Principal Component method with varimax rotation`
	`library (psych)`
	`principal(correlation,nfactors=3, rotate="varimax")`

Output	(see below)

```
Principal Components Analysis
Call: principal(r = correlation, nfactors = 3, rotate = "varimax")
Standardized loadings (pattern matrix) based upon correlation matrix
      PC3   PC1   PC2   h2   u2 com
Q1   0.66  0.39  0.01 0.58 0.42 1.6
Q2  -0.01  0.35  0.68 0.58 0.42 1.5
Q3   0.25  0.05  0.76 0.64 0.36 1.2
Q4  -0.02  0.77  0.18 0.63 0.37 1.1
Q5   0.38  0.73  0.02 0.68 0.32 1.5
Q6   0.60  0.18  0.23 0.45 0.55 1.5
Q7   0.65 -0.15  0.28 0.53 0.47 1.5
Q8   0.22  0.61  0.18 0.46 0.54 1.4
Q9   0.75  0.30 -0.29 0.74 0.26 1.6
Q10  0.49  0.15  0.32 0.37 0.63 1.9
                       PC3  PC1  PC2
SS loadings           2.29 1.95 1.42
Proportion Var        0.23 0.19 0.14
Cumulative Var        0.23 0.42 0.57
Proportion Explained  0.40 0.34 0.25
Cumulative Proportion 0.40 0.75 1.00
Mean item complexity = 1.5
Test of the hypothesis that 3 components are sufficient.
The root mean square of the residuals (RMSR) is 0.1
Fit based upon off diagonal values = 0.87
```

Factor 1: Q1, Q6, Q7, Q9, Q10 - Research Tools.
Factor 2: Q4, Q5, Q8 - Research Methods.
Factor 3: Q2, Q3 - Research Productivity.

Quality of the Factor Model

Beyond the RMSR mentioned above, a technique widely used in sociology or psychology is the reliability (Damásio, 2012). The reliability of a factor structure may be obtained by several criteria. Among other criteria presented in the literature, the calculation of the level of internal consistency by Cronbach's alpha (α) is the most used method in cross-sectional studies - when measurements are performed in a single moment (Sijtsma, 2009).

Cronbach's alpha coefficient measures the degree to which the items in an array of data are correlated. Generally, the obtained index varies between 0 and 1. A commonly accepted rule for describing internal consistency using Cronbach's alpha is:

Table 20. Python language: Principal Component method with varimax rotation (with R)

	In Python

Code	`### Principal Component method with varimax rotation (with R)`
	`import rpy2 as rpy2`
	`from rpy2.robjects.packages import importr`
	`import rpy2.robjects as ro`
	`import pandas.rpy.common as com`
	`# Changing R's directory to where is the data`
	`ro.r('setwd("C:/Users/Rui Sarmento/Documents/Livro Cybertech/Dados e Code")')`
	`# Reading the data with R`
	`ro.r('data_df <- read.csv("data.csv",sep=";")')`
	`# Retrieving the correlation matrix of the survey answers`
	`ro.r('correlation <- cor(data_df[,paste("Q",1:10,sep="")], method="spearman")')`
	`# Calling function principal with varimax rotation (with R)`
	`print(ro.r('principal(correlation,nfactors=3, rotate="varimax")'))`

Output	`Principal Components Analysis`
	`Call: principal(r = correlation, nfactors = 3, rotate = "varimax")`
	`Standardized loadings (pattern matrix) based upon correlation matrix`
	`      PC3   PC1   PC2   h2   u2 com`
	`Q1    0.66  0.39  0.01 0.58 0.42 1.6`
	`Q2   -0.01  0.35  0.68 0.58 0.42 1.5`
	`Q3    0.25  0.05  0.76 0.64 0.36 1.2`
	`Q4   -0.02  0.77  0.18 0.63 0.37 1.1`
	`Q5    0.38  0.73  0.02 0.68 0.32 1.5`
	`Q6    0.60  0.18  0.23 0.45 0.55 1.5`
	`Q7    0.65 -0.15  0.28 0.53 0.47 1.5`
	`Q8    0.22  0.61  0.18 0.46 0.54 1.4`
	`Q9    0.75  0.30 -0.29 0.74 0.26 1.6`
	`Q10   0.49  0.15  0.32 0.37 0.63 1.9`
	`                       PC3  PC1  PC2`
	`SS loadings            2.29 1.95 1.42`
	`Proportion Var         0.23 0.19 0.14`
	`Cumulative Var         0.23 0.42 0.57`
	`Proportion Explained   0.40 0.34 0.25`
	`Cumulative Proportion  0.40 0.75 1.00`
	`Mean item complexity = 1.5`
	`Test of the hypothesis that 3 components are sufficient.`
	`The root mean square of the residuals (RMSR) is 0.1`
	`Fit based upon off diagonal values = 0.87`

- 0 to 0.49 unacceptable.
- 0.50 to 0.59 poor
- 0.60 to 0.69 questionable.
- 0.70 to 0.79 acceptable.
- 0.80 to 0.89 good.
- From 0.9 to 1 excellent.

Table 21. R language: Principal Component method with varimax rotation (check the same results)

	In R

Code	`### Principal Component method`
	`library (psych)`
	`principal(new.correlation,nfactors=3, rotate="varimax")`

Output	`Principal Components Analysis`
	`Call: principal(r = new.correlation, nfactors = 3, rotate = "varimax")`
	`Standardized loadings (pattern matrix) based upon correlation matrix`
	`     PC3   PC1   PC2   h2   u2 com`
	`Q1  0.70  0.33  0.15 0.63 0.37 1.5`
	`Q2 -0.07  0.40  0.69 0.64 0.36 1.6`
	`Q3  0.12  0.05  0.81 0.67 0.33 1.1`
	`Q4  0.08  0.80  0.16 0.67 0.33 1.1`
	`Q5  0.48  0.66  0.09 0.68 0.32 1.9`
	`Q7  0.62 -0.31  0.46 0.68 0.32 2.4`
	`Q9  0.87  0.16 -0.11 0.79 0.21 1.1`
	`                       PC3  PC1  PC2`
	`SS loadings           1.88 1.47 1.40`
	`Proportion Var        0.27 0.21 0.20`
	`Cumulative Var        0.27 0.48 0.68`
	`Proportion Explained  0.40 0.31 0.30`
	`Cumulative Proportion 0.40 0.70 1.00`
	`Mean item complexity = 1.5`
	`Test of the hypothesis that 3 components are sufficient.`
	`The root mean square of the residuals (RMSR) is 0.12`
	`Fit based upon off diagonal values = 0.82`

Cronbach's alpha is influenced by the correlation values of the items and the number of evaluated items. Therefore, factors with a few items tend to have lower Cronbach's alpha and a matrix with high inter-item correlations tend to have a high Cronbach's alpha.

Remembering the retained factor of the previous subsection:

Factor 1: Q1, Q6, Q7, Q9, Q10.
Factor 2: Q4, Q5, Q8.
Factor 3: Q2, Q3.

The reliability analysis for the first factor is present in Table 23 and Table 24.

Regarding the 1st factor, the results show that $\alpha = 0.71$ (raw_alpha), i.e., an acceptable value. The values of the "reliability if an item is dropped" show a lower alpha value for all variables of this factor. This means that all of them are contributing positively to the internal consistency of the factor. Hence, it can be concluded that the first factor is well defined.

The analysis of the second factor is present in Table 25 and Table 26.

Table 22. Python language: Principal Component method with varimax rotation (check the same results)

<table>
<tr><th></th><th>In Python</th></tr>
<tr><td>Code</td><td>

```
### Principal Component method
# New correlation matrix
ro.r('new.correlation<-correlation[!(colnames(correlation) %in%
c("Q6", "Q8", "Q10")),!(rownames(correlation) %in% c("Q6", "Q8",
"Q10"))]')
print(ro.r('principal(new.correlation,nfactors=3, rotate="varimax")'))
```

</td></tr>
<tr><td>Output</td><td>

```
Principal Components Analysis
Call: principal(r = new.correlation, nfactors = 3, rotate = "varimax")
Standardized loadings (pattern matrix) based upon correlation matrix
      PC3   PC1   PC2   h2   u2 com
Q1  0.70  0.33  0.15 0.63 0.37 1.5
Q2 -0.07  0.40  0.69 0.64 0.36 1.6
Q3  0.12  0.05  0.81 0.67 0.33 1.1
Q4  0.08  0.80  0.16 0.67 0.33 1.1
Q5  0.48  0.66  0.09 0.68 0.32 1.9
Q7  0.62 -0.31  0.46 0.68 0.32 2.4
Q9  0.87  0.16 -0.11 0.79 0.21 1.1
                       PC3  PC1  PC2
SS loadings           1.88 1.47 1.40
Proportion Var        0.27 0.21 0.20
Cumulative Var        0.27 0.48 0.68
Proportion Explained  0.40 0.31 0.30
Cumulative Proportion 0.40 0.70 1.00
Mean item complexity = 1.5
Test of the hypothesis that 3 components are sufficient.
The root mean square of the residuals (RMSR) is 0.12
Fit based upon off diagonal values = 0.82
```

</td></tr>
</table>

Regarding the second factor, an alpha value of 0.66 is presented. Additionally, if some item/variable of this factor is dropped, the value of alpha decreases. Thus, it can be concluded that second factor is well defined.

The analysis of the third factor is present in Table 27 and Table 28.

Finally, the third factor has an alpha value of 0.51. Despite a little value of reliability, if some item/variable is dropped, the alpha values decreases. Again, it can be concluded that the third factor is also well defined.

Factor Analysis vs. Principal Component Analysis

In factor analysis, the different assumption about the communalities is reflected in a different correlation matrix as compared to the one used in the principal component analysis. Since in principal component analysis all communalities are initially 1, the diagonal of the correlation matrix only contains unities. In factor analysis, the initial communalities are not assumed to be 1. They are estimated (most frequently) by taking the squared multiple correlations

Table 23. R language: reliability analysis for the first factor

	In R

Code	`### Internal consistency`
	`# PC1 (Q1, Q6, Q7, Q9, Q10)`
	`library (psych)`
	`alpha(survey[c("Q1", "Q6", "Q7", "Q9", "Q10")])`

Output	`Reliability analysis`
	`Call: alpha(x = survey[c("Q1", "Q6", "Q7", "Q9", "Q10")])`
	`raw_alpha std.alpha G6(smc) average_r S/N ase mean sd`
	`0.71 0.71 0.67 0.32 2.4 0.053 3.7 0.76`
	`lower alpha upper 95% confidence boundaries`
	`0.61 0.71 0.81`
	`Reliability if an item is dropped:`
	`    raw_alpha std.alpha G6(smc) average_r S/N alpha se`
	`Q1       0.62      0.62    0.55      0.29 1.6    0.069`
	`Q6       0.66      0.66    0.61      0.33 2.0    0.064`
	`Q7       0.69      0.68    0.64      0.35 2.2    0.062`
	`Q9       0.63      0.63    0.57      0.30 1.7    0.067`
	`Q10      0.69      0.68    0.64      0.35 2.2    0.061`
	`Item statistics`
	`      n raw.r std.r r.cor r.drop mean   sd`
	`Q1  200  0.76  0.74  0.68   0.56  3.5 1.21`
	`Q6  200  0.67  0.67  0.54   0.45  3.4 1.08`
	`Q7  200  0.60  0.63  0.47   0.40  4.0 0.95`
	`Q9  200  0.73  0.72  0.63   0.52  3.8 1.19`
	`Q10 200  0.63  0.63  0.47   0.39  3.9 1.13`
	`Non missing response frequency for each item`
	`       1    2    3    4    5 miss`
	`Q1  0.08 0.14 0.20 0.34 0.24    0`
	`Q6  0.06 0.16 0.26 0.40 0.13    0`
	`Q7  0.02 0.06 0.18 0.40 0.35    0`
	`Q9  0.05 0.10 0.19 0.28 0.38    0`
	`Q10 0.04 0.10 0.14 0.34 0.37    0`

Table 24. Python language: reliability analysis for the first factor

	In Python

Code	`### Internal consistency`
	`# Cronbach alphas function`
	`def CronbachAlpha(itemscores):`
	`itemscores = np.asarray(itemscores)`
	`itemvars = itemscores.var(axis=1, ddof=1)`
	`tscores = itemscores.sum(axis=0)`
	`nitems = len(itemscores)`
	`return nitems / (nitems-1.) * (1 - itemvars.sum() / tscores.`
	`var(ddof=1))`
	`# PC1 (Q1, Q6, Q7, Q9, Q10)`
	`CronbachAlpha(np.matrix(survey_data[[0,5,6,8,9]].transpose()))`

Output	`...: CronbachAlpha(np.matrix(survey_data[[0,5,6,8,9]].transpose()))` `Out[27]: 0.70830804374937695`

Table 25. R language: reliability analysis for the second factor

	In R

Code	`### Internal consistency`
	`# PC2 (Q4, Q5, Q8)`
	`library (psych)`
	`alpha(survey[c("Q4", "Q5", "Q8")])`

Output	`Reliability analysis`
	`Call: alpha(x = survey [c("Q4", "Q5", "Q8")])`
	`raw_alpha std.alpha G6(smc) average_r S/N ase mean sd`
	`0.66 0.66 0.57 0.39 1.9 0.077 3.3 0.85`
	`lower alpha upper 95% confidence boundaries`
	`0.51 0.66 0.81`
	`Reliability if an item is dropped:`
	`raw_alpha std.alpha G6(smc) average_r S/N alpha se`
	`Q4 0.60 0.61 0.43 0.43 1.53 0.11`
	`Q5 0.47 0.48 0.31 0.31 0.91 0.12`
	`Q8 0.60 0.61 0.43 0.43 1.53 0.11`
	`Item statistics`
	`n raw.r std.r r.cor r.drop mean sd`
	`Q4 200 0.72 0.75 0.55 0.44 3.2 0.96`
	`Q5 200 0.82 0.81 0.66 0.53 3.6 1.19`
	`Q8 200 0.77 0.75 0.55 0.45 3.1 1.13`
	`Non missing response frequency for each item`
	`1 2 3 4 5 miss`
	`Q4 0.03 0.20 0.40 0.28 0.09 0`
	`Q5 0.06 0.16 0.18 0.34 0.26 0`
	`Q8 0.08 0.26 0.24 0.33 0.10 0`

Table 26. Python language: reliability analysis for the second factor

	In Python

Code	`### Internal consistency`
	`# PC2 (Q4, Q5, Q8)`
	`CronbachAlpha(np.matrix(survey_data[[3,4,7]].transpose()))`

Output	`CronbachAlpha(np.matrix(survey_data[[3,4,7]].transpose()))` `Out[28]: 0.65970835814273876`

of the variables with other variables (Rietveld and Van Hout, 1993). These estimated communalities are then represented on the diagonal of the correlation matrix, from which the eigenvalues will be determined, and the factors will be retained. After extraction of the factors, new communalities can be calculated, which will be represented in a reproduced correlation matrix (Kootstra, 2004).

The difference between factor analysis and principal component analysis is crucial in interpreting the factor loadings: by squaring the factor loading

Table 27. R language: reliability analysis for the third factor

	In R

Code	`### Internal consistency`
	`# PC3 (Q2, Q3)`
	`library (psych)`
	`alpha(survey[c("Q2", "Q3")])`

Output	`Reliability analysis`
	`Call: alpha(x = survey [c("Q2", "Q3")])`
	`raw_alpha std.alpha G6(smc) average_r S/N ase mean sd`
	`0.51 0.52 0.35 0.35 1.1 0.12 3.9 0.77`
	`lower alpha upper 95% confidence boundaries`
	`0.27 0.51 0.74`
	`Reliability if an item is dropped:`
	`   raw_alpha std.alpha G6(smc) average_r S/N alpha se`
	`Q2      0.35      0.35    0.12      0.35  NA       NA`
	`Q3      0.35      0.35    0.12      0.35  NA       NA`
	`Item statistics`
	`     n raw.r std.r r.cor r.drop mean   sd`
	`Q2 200  0.78  0.82  0.48   0.35  3.9 0.83`
	`Q3 200  0.86  0.82  0.48   0.35  3.9 1.04`
	`Non missing response frequency for each item`
	`      1    2    3    4    5 miss`
	`Q2 0.00 0.06 0.21 0.48 0.25    0`
	`Q3 0.02 0.10 0.18 0.39 0.31    0`

Table 28. Python language: reliability analysis for the third factor

	In Python

Code	`### Internal consistency`
	`# PC3 (Q2, Q3)`
	`CronbachAlpha(np.matrix(survey_data[[1,2]].transpose()))`

Output	`CronbachAlpha(np.matrix(survey_data[[1,2]].transpose()))` `Out[29]: 0.50738200972962688`

of a variable, the amount of variance accounted by that variable is obtained. However, in factor analysis, it is already initially assumed that the variables do not account for 100% of the variance. Thus, as Rietveld & Van Hout (1993) state, "although the loading patterns of the factors extracted by the two methods do not differ substantially, their respective amounts of explained variance does!"

CONCLUSION

Factor analysis is an important task to execute a quantity reduction of variables. This is particularly useful when there are many variables in our data.

Thus, this is an important research area that should be carefully studied by the data analyst. The central concepts that were presented in this chapter are:

- Bartlett Sphericity test.
- KMO measure.
- Retained factors:
    - Kaiser criterion,
    - Scree plot,
    - Variance explained criteria.
- Factor analysis methods:
    - Principal component analysis,
    - Principal axis of factor extraction,
    - Maximum likelihood.
- Factor rotations.
- Internal consistency with Cronbach's alpha.

REFERENCES

Abdi, H. (2003). Factor rotations in factor analyses. In Encyclopedia for Research Methods for the Social Sciences (pp. 792-795). Thousand Oaks, CA: Sage.

Cattell, R. B. (1966). The scree test for the number of factors. *Multivariate Behavioral Research*, *1*(2), 245–276. doi:10.1207/s15327906mbr0102_10 PMID:26828106

Damásio, B. F. (2012). Uso da análise fatorial exploratória em psicologia. *Avaliação psicológica*, *11*(2), 213-228.

Hayton, J. C., Allen, D. G., & Scarpello, V. (2004). Factor Retention Decisions in Exploratory Factor Analysis: A Tutorial on Parallel Analysis. *Organizational Research Methods*, *7*(2), 191–205. doi:10.1177/1094428104263675

Kaiser, H. F. (1958). The varimax criterion for analytic rotation in factor analysis. *Psychometrika*, *23*(3), 187–200. doi:10.1007/BF02289233

Kootstra, G. (2004). *Exploratory factor analysis*. Unpublished Paper. Retrieved from http://www.let.rug.nl/nerbonne/teach/ rema-stats-meth-seminar/ Factor-Analysis-Kootstra-04.PDF

Ledesma, R. D., & Valero-Mora, P. (2007). Determining the Number of Factors to Retain in EFA: An easy-to-use computer program for carrying out Parallel Analysis. *Practical Assessment, Research & Evaluation*, *12*(2).

Marôco, J. (2011). Análise Estatística com o SPSS Statistics (5th ed.). Pero Pinheiro.

Rietveld, T., & Van Hout, R. (1993). *Statistical techniques for the study of language and language behaviour*. Walter de Gruyter. doi:10.1515/9783110871609

Sijtsma, K. (2009). On the use, the misuse, and the very limited usefulness of Cronbachs alpha. *Psychometrika*, *74*(1), 107–120. doi:10.1007/s11336-008-9101-0 PMID:20037639

Chapter 8
Clusters

INTRODUCTION

Cluster analysis was originated in anthropology by Driver and Kroeber in 1932. It is the task of grouping a set of objects in such a way that objects in the same group or cluster are more similar to each other than to those in other groups or clusters. It is a common technique for statistical data analysis.

Cluster analysis can be achieved by various algorithms that might differ significantly. Modern notions of clusters include groups with small distances among the cluster members, dense areas of the data space, intervals or particular statistical distributions. Therefore, cluster analysis as such is not a trivial task. It is an interactive multi-objective optimization that involves trial and error. It is also often necessary to modify data preprocessing and model or algorithms parameters until the result achieves the desired characteristics.

Therefore, in cluster analysis, the clustering of subjects or variables are made from similarity measures or dissimilarity (distance) between two subjects initially, and later between two clusters. These groups can be done using hierarchical or non-hierarchical techniques.

R VS. PYTHON

In this chapter, cluster analysis will be, once more, also conducted also in R language and Python. First, a brief theoretical summary will be necessary for the reader to understand the concepts regarding clusterization.

DOI: 10.4018/978-1-68318-016-6.ch008

Similarity and Dissimilarity

The identification of natural clusters of subjects or variables requires that the similarity between these have to be measured explicitly. There are several similarity measures (or proximity) or dissimilarity (or distance) that can be used depending on the variable type (interval, frequency or nominal). In cluster analysis, the most common measures are given below.

Euclidean Distance

This is the distance between two points (p, q) in any dimension of the space and is the most common use of distance. When data is dense or continuous, this is the best proximity measure. Euclidean distance measure is given by:

$$d\left(p,q\right)=\sqrt{\sum_{k=1}^{n}\left(p_k-q_k\right)^2}$$

Minkowski Distance

This is a metric in a normed vector space, which can be considered a generalization of the Euclidean distance. The Minkowski distance measured between two points (p, q) is given by:

$$d\left(p,q\right)=\left(\sum_{k=1}^{n}\left|p_k-q_k\right|^c\right)^{\frac{1}{c}}$$

with $c=1$ and $c=2$, the Minkowski metric becomes equal to the Manhattan and Euclidean metrics respectively.

Cosine Similarity

It is often used when comparing two documents against each other. It measures the angle between two vectors. If the value is zero, the angle between the two vectors is 90 degrees, and they share no terms. If the value is one, the two vectors are the same except for magnitude. Given two vectors of attributes, *u* and *v*, the cosine similarity, $\cos\theta$, is represented as

$$\cos\theta = \frac{u \cdot v}{uv} = \frac{\sum_{i=1}^{n} u_i v_i}{\sqrt{\sum_{i=1}^{n} u_i^2}\sqrt{\sum_{i=1}^{n} v_i^2}}$$

where u_i and v_i are components of vector u and v, respectively.

Jaccard Similarity

This is a standard index for binary variables. It is defined as the quotient between the intersection and the union of the pairwise compared variables between two objects.

The Jaccard distance between the objects i and j is given by

$$d(i,j) = \frac{M_{11}}{M_{01} + M_{10} + M_{11}}$$

M_{11} represents the total number of attributes where both data objects have a 1; and M_{10}, M_{01} represent the total number of attributes where one data object has a 1, and the other has a 0. The total matching attributes are then divided by the total non-matching attributes, plus the matching ones. A perfect similarity score would then be 1.

Similarity Measures for Variables

When cluster analysis aims to group variables (and not subjects or items), the appropriate similarity measures are the sample correlation coefficients. In case of continuous variables, Pearson correlation coefficient is the most suitable. For ordinal variables the Spearman correlation coefficient should be used. Finally, for nominal variables, the reader should use the phi coefficient, $\phi = \sqrt{\frac{X^2}{N}}$, where X^2 is the chi-square statistic.

Regarding the data used in the clusters study, only researchers with the largest number of publications in 2015 or 2016 were considered. Cluster analysis could be executed with any sample. However, to be easier to explain, this filtering of researchers was performed. Moreover, the Euclidean distance was chosen. The results are presented in Table 1 and Table 2.

Table 1. R language: Euclidean distance matrix

In R

Code

```
### Euclidean distance matrix
# Filtering some data from 2015
newdata<-data[data$Year>=2015,]
# Distance´s matrix
d <- dist(as.matrix(newdata[,paste("Q", 1:10, sep="")],
method="euclidean"))
d
```

Output

	3	22	50	80	94	95	99	(…)
22	9.746794							
50	9.000000	5.477226						
80	1.732051	9.486833	9.055385					
94	2.645751	9.380832	9.055385	2.449490				
95	0.000000	9.746794	9.000000	1.732051	2.645751			
99	2.236068	9.165151	8.602325	2.449490	2.828427	2.236068		
105	1.732051	10.295630	9.273618	2.000000	3.162278	1.732051	2.449490	
107	8.185353	5.830952	4.472136	8.717798	8.602325	8.185353	7.874008	(…)
111	8.306624	6.324555	5.099020	8.602325	8.485281	8.306624	8.124038	(…)
(…)	(…)	(…)	(…)	(…)	(…)	(…)	(…)	(…)

The previous outputs show an excerpt from the distances matrix calculated by Euclidean distance method. In this excerpt, it is possible to find both items with distance around 9 (e.g. items 3 and 22) and items with distance very near to 0 or 1 (e.g. items 3 and 95). Thus, these preliminary results of this study suggest that at least two clusters exist. Nonetheless, further investigation has to be held.

Hierarchical Clustering

Hierarchical techniques appeal to successive steps of aggregation of the considered subjects, individually. Thus, given a set of N items to be clustered, and a $N \times N$ distance (or similarity) matrix, the primary process of hierarchical clustering is:

1. Start by assigning each item to its cluster, so that if it has N items, it now has N clusters, each containing just one item. Let the distances (similarities) between the clusters equal the distances (similarities) between items they contain.

Table 2. Python language: Euclidean distance matrix

	In Python

Code	`### Euclidean distance matrix`
	`# Import libraries`
	`import scipy.spatial.distance as sp`
	`# Filtering some data from 2015`
	`new_data = data[data['Year'] >= 2015]`
	`# Filtering survey data`
	`survey_data = new_data.ix[:,7:17]`
	`# Calculate distance`
	`X = sp.pdist(survey_data, 'euclidean')`
	`X`

Output	Out[34]: array([9.74679434, 9. , 1.73205081, 2.64575131, 0. , 2.23606798, 1.73205081, 8.18535277, 8.30662386, 1. , 8.71779789, 1.41421356, 9.21954446, 1.73205081, 1.73205081, 1.73205081, 9.32737905, 1.41421356, 5.47722558, 9.48683298, 9.38083152, 9.74679434, 9.16515139, 10.29563014, 5.83095189, 6.32455532, 9.48683298, 5.91607978, 10.04987562, 6.32455532, 9.59166305, 10.19803903, 10.19803903, 7.87400787, 10.04987562, 9.05538514, 9.05538514, 9. , 8.60232527, 9.2736185 , 4.47213595, 5.09901951, 8.71779789, 3.31662479, 9.11043358, 3.16227766, 8.83176087, 9.69535971, 9.69535971, 6.92820323, 9.53939201, 2.44948974, 1.73205081, 2.44948974, 2. , 8.71779789, 8.60232527, 2. , 9.11043358, 2.23606798, 9.2736185 , 2.44948974, 1.41421356, 1.41421356, 9.89949494, 1.73205081, 2.64575131, 2.82842712, 3.16227766, 8.60232527, 8.48528137, 3.16227766, 9.11043358, 3.60555128, 9.16515139, 2.82842712, 2.82842712, 2.82842712, 9.48683298, 2.23606798, 2.23606798, 1.73205081, 8.18535277, 8.30662386, 1. , 8.71779789, 1.41421356, 9.21954446, 1.73205081, 1.73205081, 1.73205081, 9.32737905, 1.41421356, 2.44948974, 7.87400787, 8.1240384 , 2.82842712, 8.42614977, 3. , 8.94427191, 1.41421356, 2. , 2. , 9.16515139, 1.73205081, 8.48528137, 8.48528137, 2.44948974, 8.88819442, 1.73205081, 9.38083152, 2. , 1.41421356, 1.41421356, 9.59166305, 1.73205081, 3.16227766, 8. , 2.64575131, 8.30662386, 4.24264069, 7.87400787, 9.05538514, 9.05538514, 4. , 8.77496439, 8.1240384 , 3. , 8.18535277, 3.16227766, 8. , 9.16515139, 9.16515139, 3.16227766, 9. , 8.54400375, 1.73205081, 8.94427191, 2.44948974, 2.44948974, (...)])

2. Find the closest (most similar) pair of clusters and merge them into a single cluster, so that now it has one less cluster.
3. Compute distances (similarities) between the new cluster and each of the old clusters.
4. Repeat steps 2 and 3 until all items are clustered into a single cluster of size N.

Hierarchical methods of clusters mostly differ in how these distances (in step 3) are calculated. The methods most frequently used are as follows.

Single-Linkage Clustering

Single linkage (also called connectedness or minimum method) is one of the simplest agglomerative hierarchical clustering methods. In single linkage, the distance between groups is defined as the distance between the closest pair of objects, where only pairs consisting of one object from each group are considered.

In single linkage method, $D(r,s)$ is computed as $D(r,s) = \min(d(i,j))$, where i is in cluster r and j is in cluster s. Thus, the distance between two clusters is given by the value of the shortest link between the clusters.

Complete Linkage Clustering

In complete linkage (also called farthest neighbor), the clustering method is the opposite of single linkage. The distance between groups is defined as the distance between the most distant pair of objects, one from each group.

In complete linkage method, $D(r,s)$ is computed as $D(r,s) = \max(d(i,j))$, where i is in cluster r and object j is in cluster s. Thus, the distance between two clusters is given by the value of the longest link between clusters.

Average Group Linkage

With average group linkage, the groups formed are represented by their mean values for each variable (i.e., their mean vector and inter-group distance is defined regarding the distance between two such mean vectors).

In average group linkage method, the two clusters, r, and s, are merged such that the average pairwise distance within the newly formed cluster is minimum. Suppose the new cluster formed by combining clusters r and s is labeled as t. Then the distance between clusters r and s, $D(r,s)$, is computed as $D(r,s) = Average(d(i,j))$, where observations i and j are in cluster t, the cluster formed by merging clusters r and s.

At each stage of hierarchical clustering, the r and s clusters for which $D(r,s)$ is minimum, are merged. In this case, those two clusters are merged such that the newly formed cluster, on average, will have minimum pairwise distances between the points.

Average Linkage within Groups

Average linkage within groups is a technique of cluster analysis in which clusters are combined in order to minimize the average distance between all individuals or cases in the resulting cluster. Also, the distance between two clusters is defined as the average distance between all possible pairs of individuals in the cluster that would result if they were combined.

Centroid Clustering

A cluster centroid is the middle point of a cluster. A centroid is a vector containing one number for each variable, where each number is the mean of a variable for the observations in that cluster.

The reader can use the centroid as a measure of cluster location. For a particular cluster, the average distance from the centroid is the average of the distances between observations and the centroid. The maximum distance from the centroid is the maximum of these distances.

Ward Method

It is an alternative approach for performing cluster analysis. Essentially, it looks at cluster analysis as an analysis of variance problem, instead of using distance metrics or measures of association.

This method involves an agglomerative clustering algorithm. It will start out at the leaves and work its way to the trunk. It looks for groups of leaves that it forms into branches, the branches into limbs and eventually into the trunk. Ward's method starts out with *n* clusters of size 1 and continues until all the observations are included into one cluster.

This method is the most appropriate for quantitative variables and not binary variables.

As there are several available methods, the existence of advantages and disadvantages in using each one of them is visible. Since the "best" method of performing hierarchical clustering does not exist, some authors (Marôco, 2011) suggest the use of various methods simultaneously. Hence, if all methods produce similarly interpretable solutions, it is possible to conclude that data matrix has natural groupings.

Returning to the data of this book, two methods are applied: single-linkage clustering and complete-linkage clustering. The respective dendrograms are shown in Table 3 and Table 4.

The analysis of dendrograms suggests the existence of two clusters.

Table 3. R language: hierarchical clustering

<table>
<tr><th></th><th>In R</th></tr>
<tr><td>Code</td><td>

```
### Hierarchical clustering
a) # Single-linkage clustering (method= "single")
hc <- hclust(d, method = "single")
plot(hc)

b) # Complete-linkage clustering (method= "complete" by default)
hc <- hclust(d)
plot(hc)
# Draws rectangles around the branches of a dendrogram highlighting
the corresponding clusters
rect.hclust(hc, k = 2)
```

</td></tr>
<tr><td>Output</td><td>

a) Dendrogram with single linkage method in R:

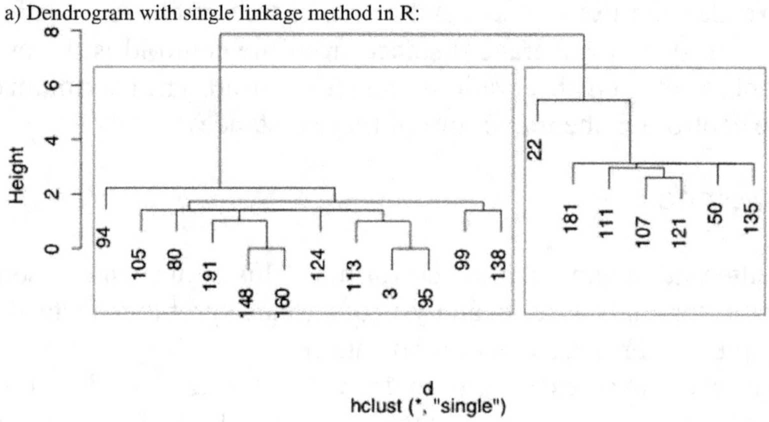

b) Dendrogram with complete linkage method in R:

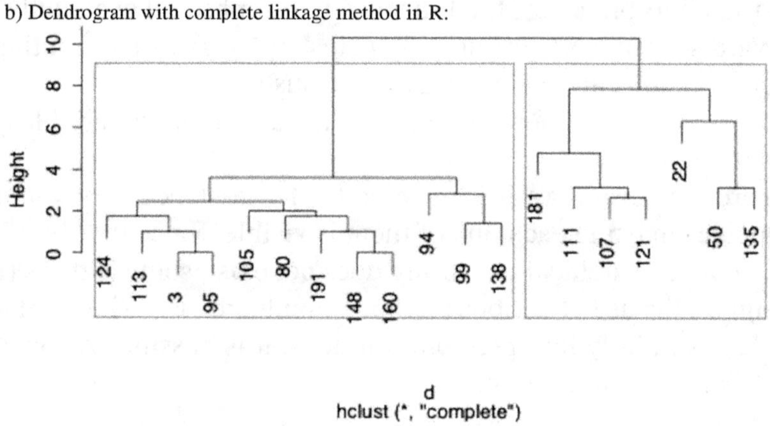

</td></tr>
</table>

Clusters

Table 4. Python language: hierarchical clustering

	In Python

Code	```### Hierarchical clustering
	from matplotlib import pyplot as plt
	from scipy.cluster.hierarchy import dendrogram, linkage
	Z_single = linkage(X, 'single') # Single linkage
	Z_complete = linkage(X, 'complete') # Complete linkage
	
	a) # Single-linkage clustering (method= "single")
	# Calculate full dendrogram
	plt.figure(figsize=(25, 10))
	plt.title('Hierarchical Clustering Dendrogram')
	plt.xlabel('sample index')
	plt.ylabel('distance')
	dendrogram(Z_single,
	leaf_rotation=90., # rotates the x axis labels
	leaf_font_size=8., # font size for the x axis labels
	)
	plt.show()
	
	b) # Complete-linkage clustering (method= "complete")
	plt.figure(figsize=(25, 10))
	plt.title('Hierarchical Clustering Dendrogram')
	plt.xlabel('sample index')
	plt.ylabel('distance')
	dendrogram(Z_complete,
	leaf_rotation=90., # rotates the x axis labels
	leaf_font_size=8., # font size for the x axis labels
	)
	plt.show()```

Output	a) Dendrogram with single linkage method in Python: 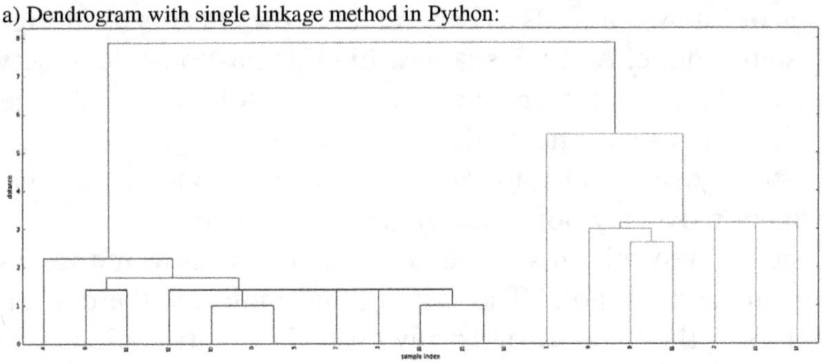 b) Dendrogram with complete linkage method in Python: 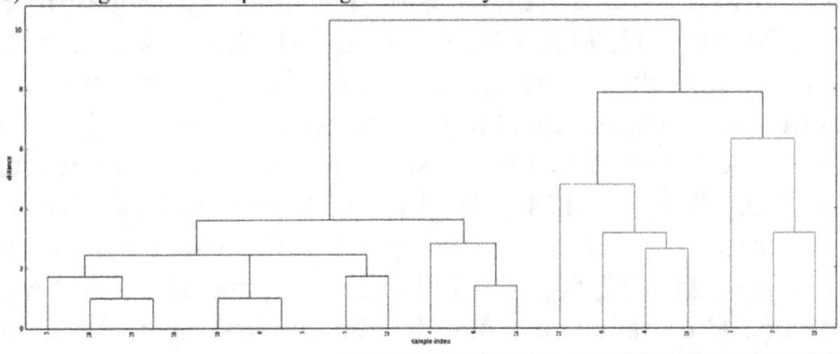

Non-Hierarchical Cluster Analysis

Non-hierarchical clustering methods are intended in grouping items (and not variables) in a set of clusters whose number is defined a-priori. These methods quickly apply to arrays of large data because it is not necessary to calculate and store a new dissimilarity matrix in each step of the algorithm.

There are various non-hierarchical methods that differ primarily in the way it unfolds the first aggregation of items in clusters and how the new distances between the centroids of the clusters and the item are calculated. One of the standard methods in most statistical software is the K-means.

K-Means

The procedure follows a straightforward and easy way to classify a given data set with a specified number of clusters (assume *k* clusters) fixed a-priori. The main idea is to determine *k* centers, one for each cluster. These centers should be placed in a cunning way because of different location causes a different result. Thus, the better choice is to place them far away from each other, as much as possible. The next step is to take each point belonging to a given data set and associate it to the nearest center. When no point is pending, the first phase is completed, and an early group is done. At this point, the re-calculation of *k* new centroids as barycenter of the clusters resulting from the previous step is done. After this, a new binding has to be done between the same data set points and the nearest new center. A loop has been generated. As a result of this loop, the reader may notice that *k* centers change their location, step by step, until no more changes are done or, in other words, centers do not move anymore. See Table 5 and Table 6.

In R, the first ten columns of the previous output show the values of each researcher in each variable. The last column indicates the cluster of each researcher. In Python, the second array gives the identification of each cluster. Thus, the researchers with the numbers 3, 80, 94, 95, 99, 105, 113, 124, 138, 148, 160, and 191 are in one cluster. The other cluster has the researchers number 22, 50, 107, 111, 121, 135, and 181. Since two clusters and the researchers of each cluster are known, now it is important to identify what separates the two clusters. Checking the answers of these analyzed researchers, it is possible to verify that the first cluster (constituted by researchers 3, 80, 94, 95, 99, 105, 113, 124, 138, 148, 160, and 191) gave very positive answers to the most questions. On the opposite, the researchers of the second cluster (composed by 22, 50, 107, 111, 121, 135, and 181) had lower values in the answers. Thus, it can be concluded that the first cluster is characterized

Clusters

Table 5. R language: clustering with k-means

In R

Code

```
### Clustering with k-means
# Selecting data
newdata <- newdata[,paste("Q", 1:10, sep="")]
# K-means cluster analysis
fit <- kmeans(newdata, 2)
# Append cluster assignment
newdata <- data.frame(newdata, fit$cluster)
```

Output

```
    Q1 Q2 Q3 Q4 Q5 Q6 Q7 Q8 Q9 Q10 fit.cluster
3    5  5  5  4  5  4  5  4  5   5           1
22   1  3  5  3  1  1  1  5  1   1           2
50   1  3  2  3  1  1  2  3  1   5           2
80   5  5  5  5  5  4  4  5  5   5           1
94   5  5  5  3  5  5  3  5  5   5           1
95   5  5  5  4  5  4  5  4  5   5           1
99   4  4  5  4  4  5  5  5  5   5           1
105  5  5  4  5  5  5  5  4  5   5           1
107  2  2  2  2  2  2  4  2  2   2           2
111  2  2  2  3  2  2  2  1  4   2           2
113  5  5  5  4  5  3  5  4  5   5           1
121  1  3  2  3  1  2  3  1  2   3           2
124  5  5  5  5  5  4  5  3  5   5           1
135  2  2  1  3  1  1  1  2  3   4           2
138  5  4  5  4  4  5  5  4  5   5           1
148  5  5  5  5  5  5  5  5  5   5           1
160  5  5  5  5  5  5  5  5  5   5           1
181  3  2  1  1  1  2  3  1  5   1           2
191  5  5  5  4  5  5  5  5  5   5           1
```

Table 6. Python language: clustering with k-means

In Python

Code

```
### Clustering with k-means
# Import modules
from scipy.cluster.vq import kmeans2
from scipy.cluster.vq import whiten
# Normalize variables values
std_survey_data = whiten(survey_data, check_finite=True)
# K-means cluster analysis
kmeans2(std_survey_data,2, iter=10)
```

Output

```
Out[37]:
(array([[ 3.04045353, 3.91578649, 3.16712161, 4.0174553, 2.74895893,
2.93608205, 3.33573599, 3.02022692, 3.4551166, 3.4551166 ],
[ 1.0601097, 1.96753804, 1.38034356, 2.38398446, 0.73125016,
1.0252985, 1.60516619, 1.46533921, 1.77691711, 1.77691711]]),
array([0, 1, 1, 0, 0, 0, 0, 0, 1, 1, 0, 1, 0, 1, 0, 0, 0, 1, 0]))
```

by having researchers with a positive opinion about tools, methods, and productivity. The second cluster is marked by having researchers with a negative view of tools, methods, and productivity.

CONCLUSION

The clustering analysis presented in this chapter allows the reader to quickly identify groups of individuals with different characteristics in a large data sample. If the reader knows the number of clusters, non-hierarchical clustering is sufficient. If not, a hierarchical analysis must be performed preliminarily.

The central concepts of this chapter are:

- Hierarchical clustering.
- Non-hierarchical clustering.
- Clustering methods:
    - Single linkage,
    - Complete linkage,
    - Average group linkage,
    - Average linkage within groups,
    - Centroid clustering,
    - Ward method,
    - K-means.

REFERENCES

Driver, H. E., & Kroeber, A. L. (1932). Quantitative expression of cultural relationships. *University of California Publications in American Archeology and Ethnology*, *31*(4), 211–256.

Marôco, J. (2011). Análise Estatística com o SPSS Statistics (5th ed.). Pero Pinheiro.

Chapter 9
Discussion and Conclusion

INTRODUCTION

After all the material we covered throughout this book, this chapter ends the book with a discussion and conclusion about the document's purpose. Thus, in this chapter, we try to clearly state the reasons why we have used the tools we chosen for the statistical analysis tasks and finally conclude the comparison between them.

DISCUSSION

As previously reported, besides focusing on the analytical tasks in this book, we provided practice procedures and examples in both R and Python languages. There are many information sources about these languages. We detailed a brief summary of both languages characteristics in the course of this book. However, the reader might be thinking what would be our preference if we had to choose between both languages. We present our considerations and points of view about this matter. We also present this in a way that appears to us as being more interesting to the reader, which is to foment a discussion about this subject, which we feel to be very, very subjective.

DOI: 10.4018/978-1-68318-016-6.ch009

R vs. Python?

The discussion of choice between R and Python when choosing a language to do data analysis is not new in the academic world. It has been discussed if a direct approach favors R in detriment of Python or a more generic approach favors Python in detriment of R. What we think about this matter is that it is mainly a subjective case of choosing between languages. This is true in other fields of study, not only in the data analysis area. The experience of the programmer with languages frequently dictates his/her choice to proceed with programming tasks. Thus, we think an experienced Python developer would naturally choose Python to do data analysis tasks even though we might believe R language, which is a specifically created language, might be easier to work with if there is no previous contact with any of these two languages.

R is a powerful programming language, used in statistical tasks, data analysis, numerical analysis and others. The main characteristics of R we previously stated in this book, favor visualization of data and the results of statistical analysis. Nonetheless, Python, as a generic programming language also provides these advantages, although this might have to be done with a little more effort from the programmer. Throughout the writing of this book, we had to develop statistical algorithms in Python because, to our knowledge, there were no module functions available to proceed with the tasks. Additionally, not less than once, we had to use R within Python to finish certain statistical analysis tasks. This is something that might have two different readings, depending on the point of view. For the experienced Python programmer, this might be seen as added freedom degrees because Python is a generic programming language. The possibility of connection to other languages and the freedom to create an original piece of code to solve the problem is indeed a dominant characteristic of universal programming languages. On the other side, for less experienced programmers, this added effort of having to use other languages, to install added packages or even to have to program additional code might be a handicap. Consequently, on the downside, those added degrees of programming freedom might not be initially suitable for everyone. We dealt with that in this book, to make the reader's life easier if this is the reader's situation.

Due to the relative proliferation of both languages, there is a high amount of sources of information, beyond the manuals of the programming languages. Users of both languages have long been discussing the advantages and disadvantages in internet forums, blogs, Webpages, pseudo-manuals, scientific publications and others. Through the point of view of the inexperienced programmer, this might be good, and his/her confidence of getting help when in difficulty is high. Nonetheless, this information is scattered and spread in such an amount of sources we feel sometimes is difficult to connect everything in a consistent result. Again, in this book, we try to aggregate all the crucial available information to proceed with data analysis with any chosen language, R, Python or even both.

CONCLUSION

In this book, we presented data analysis procedures in Python and R languages. In the first chapters, we presented several subjects, to introduce the reader to Statistics. Additionally, the reader was introduced to simple programming tasks and to provide his/her comprehension of the use of features that would be applied elsewhere in the book. Although the book was organized with a crescent complexity of materials, the reader encountered an imminently practical book with examples throughout.

Additionally, in the first chapters, we provided a brief summary of the syntax of the languages we are focusing on this book. We introduced the reader to their consoles, GUIs and IDEs, either for R or Python. We stress that, in that initial chapter, we approached just a little bit of the existing material regarding both programming languages. Nonetheless, we believe that it is possible for the reader to gather information from other sources and we tried to state them also in the initial chapters.

Following this brief introduction to programming in both languages, a chapter regarding the dataset used in this book was also presented. In this chapter, a brief summary of the variables characteristics was addressed. These variables supported all analysis tasks throughout the book.

Then, a chapter regarding descriptive analytics is also available. This chapter presented the main concepts regarding data descriptive and visualization. Inferential analysis was also performed. The main ideas regarding this area were submitted and supported by examples with both programming languages. A brief introduction to regression analysis is also presented. Then, we conclude the book with two chapters, one directed to Factor Analysis and one regarding Clusters.

Table 1 summarizes the concepts approached in each main chapter.

With all these concepts understood, the reader should be capable of performing data analysis on its own. Additionally, the reader should now be able to perform a complete analysis with R or Python languages. It should be noticed that a full statistical data analysis of any data should follow a particular workflow. The sequence of some of the chapters presented in this

Table 1. Chapter concepts summarized

Statistics	Intro to Programming	Dataset	Descriptive Analytics
• Variable, population and sample • Mean, median, mode, standard deviation, quartile an percentile • Statistic distributions o Normal distribution o Chi-square distribution o Student's t-distribution o Snedecor's F-distribution o Binomial distribution • Central limit theorem • Decision rules: p-value, error, confidence interval and tests.	Programming of: • Vectors • Dataframes • Matrices • Functions , and the installation of: • R • RStudio • Anaconda Python's Distribution	• Dataset Variables • Variable's Types • Pre-processing in R (Introduction) • Pre-processing in Python (Introduction)	• Variables Frequencies' Analysis • Measures of Central Tendency and Dispersion o Median o Mean o Standard Deviation o Max o Min o Quartiles o Outlier and missing values • Graphical representation of variables o Pie Chart o Bar Graph o Boxplot o Histogram
Inferential Analysis	**Intro to Regression**	**Factor Analysis**	**Clusters**
• Statistical tests o Shapiro-Wilk test o Kolmogorov-Smirnov test o Student's t-test o ANOVA o Levene's test o Welch correction o Games-Howell test o Tukey test o Mann-Whitney test o Kruskal-Wallis test o Chi-square test • Correlations o Pearson o Spearman o Kendall	• ANOVA for regression • Coefficient of determination • Multiple Linear Regression	• Bartlett Sphericity test • KMO measure • Retained factors o Kaiser criterion o Scree plot o Variance explained criteria • Factor analysis methods o Principal component analysis o Principal axis of factor extraction o Maximum likelihood • Factor rotations • Internal consistency with Cronbach alpha	• Hierarchical clustering • Non-hierarchical clustering • Clustering methods o Single linkage o Complete linkage o Average group linkage o Average linkage within groups o Centroid clustering o Ward method o K-means

book follows the workflow the authors of this book feel to be the most common and appropriate. A summary of this common workflow was shown in Figure 1 in the book's introduction. In the end, the choice between Python or R is entirely a choice to be done by the reader. The authors will not favor any of the languages used in this book. However, the authors are sure the reader will do just as well, either deciding to proceed with R or Python.

About the Authors

Rui Sarmento has a degree in Electrical Engineering in the Faculty of Engineering, University of Porto and a MSc in Data Analysis and Decision Support Systems in the Faculty of Economics of the University of Porto. He has worked in several areas from an international technical support center to software development companies focusing on Communications and Intranet solutions with Linux-based Enterprise Operating Systems. Finally, he has also worked for the main public transportation company in his hometown, Porto, as a Project Management engineer in the IT area. He is currently also collaborating with LIAAD (Laboratory of Artificial Intelligence and Decision Support) in INESC TEC researching on Large Social Networks Analysis and Visualization. Rui was previously published as a Contributing Author in the book *Integration of Data Mining in Business Intelligence Systems* (IGI Global, 2014).

Vera Costa has a degree in Mathematics at the Faculty of Sciences of the University of Porto and a Master Degree in Data Analysis and Decision Support Systems at the Faculty of Economics of the same university. She started to teach mathematics to high school and, after that, information systems to college students. She participated in several research projects, and she contributes with her knowledge of statistical analysis and software programming, in different application fields, such as health, politics, and transportation. Currently, she is a Ph.D. student of Transportation Systems of the MIT Portugal program.

Index

R

S

V

www.ingramcontent.com/pod-product-compliance
Lightning Source LLC
LaVergne TN
LVHW061534290926
849162LV00004B/786